# Praise for *Be[illegible]*

"In this superbly written book, Ia[illegible] knowledge of the human fossil r[illegible], Paleolithic archaeology, primate behavior, prehistoric art, as well as the workings of the human brain and our extraordinary cognitive powers, to offer a convincing scenario of how we have come to hold dominion over the earth."
—Donald Johanson, *Scientific American*

"There is no more literate anthropologist writing on human evolution today than Ian Tattersall, curator of anthropology at the American Museum of Natural History in New York. His important new book, *Becoming Human*, is at once absorbing in its details, provocative in its thoughtful speculations and delightfully informal in its style."
—*San Francisco Chronicle*

"Although many popular anthropological accounts of the human species have been written, few are as engaging as that of Ian Tattersall."
—*Natural History*

"Tattersall is both a real, working scientist and a gifted writer. His latest effort, *Becoming Human*, is an ambitious effort—are there any bigger subjects than describing what makes us?—but Tattersall meets the challenge commendably."
—*The San Diego Union-Tribune*

"Tattersall deftly tells the complex story of how we have come to be our unique selves in this, the best book on human evolution I've ever read."
—Niles Eldredge, Curator, Department of Invertebrates,
The American Museum of Natural History

"Asserting that modern humans have a creative spark that sets them apart from their predecessors, [Ian Tattersall] presents a thoughtful narrative about the symbolic representation and religious beliefs of Cro-Magnons." —*Science News*

"Devoted to a wide-ranging—and highly readable—tour of the fossil evidence of how, anatomically, we got to where we are today . . . wittily and cogently argued, and uncompromising."
—*New Scientist*

"Tattersall engages the reader in a fascinating contemplation of just how we differ from all other organisms."
—Jeffrey Schwartz, author of *The Red Ape* and *What the Bones Tell Us*

"Through an enlightening examination of the latest fossil evidence, we gain new insights into the role of competition among species, the impact of climate, and the episodic vs. gradual nature of evolutionary changes." —*Library Journal*

"A felicitous writer, Tattersall gracefully summarizes what science knows for certain about human origins and indicates areas, such as language development, where debate rages." —*Booklist*

# Becoming *Human*

ALSO BY IAN TATTERSALL

Man's Ancestors: An Introduction to Primate and Human Evolution *(1970)*

Lemur Biology *(1975) (ed. with R. W. Sussman)*

The Primates of Madagascar *(1982)*

The Myths of Human Evolution *(1982) (with Niles Eldredge)*

Encyclopedia of Human Evolution and Prehistory *(1988) (ed. with Eric Delson and J. A. Van Couvering)*

The Human Odyssey: Four Million Years of Human Evolution *(1993)*

Lemurs of Madagascar: A Field Guide *(1994) (with R. A. Mittermeier, W. R. Konstant, D. M. Meyers, R. B. Mast, and Stephen Nash)*

Primates: Lemurs, Monkeys and You *(1995)*

The Fossil Trail: How We Know What We Think We Know about Human Evolution *(1995)*

The Last Neanderthal: The Rise, Success and Mysterious Extinction of Our Closest Human Relative *(1995)*

# Becoming *Human*

## *Evolution and Human Uniqueness*

Ian Tattersall

A HARVEST BOOK

HARCOURT BRACE & COMPANY

*San Diego New York London*

Library of Congress Cataloging-in-Publication Data
Tattersall, Ian.
Becoming human: evolution and human uniqueness/
Ian Tattersall. —1st ed.
p. cm.
Includes bibliographical references and index.
ISBN 0-15-100340-8
ISBN 0-15-600653-7 (pbk.)
1. Human evolution. 2. Primates—Evolution. 3. Fossil hominids.
4. Social evolution. I. Title.
GN281.T355 1998
599.93'8—dc21 97-30434

Text set in Janson Text
Designed by Kaelin Chappell and Rich Hendel
Printed in the United States of America
First Harvest edition 1999
E D C B

FOR MY SISTER JILL

*with affection and gratitude*

# ACKNOWLEDGMENTS

I have learned from many people over the years, but here I would especially like to acknowledge my debt to my friend and colleague Niles Eldredge, who will recognize here many of the themes we have debated over fluids of various hues ever since those days at the Blarney Castle a quarter century ago. I am indebted also to my remarkable editor, Jane Isay, who conjured up the notion of a book where I didn't know one existed, and whose comments, and those of Kate Murphy, improved the final product. And thanks to Erin DeWitt for her sensitive copyediting, and to Lorie Stoopack for assistance.

CONTENTS

# Becoming *Human*

. . . . . . . . . . . .

# *Prologue*

A leisurely half hour's stroll outside the sleepy southwestern French town of Les Eyzies de Tayac, a narrow fissure penetrates deep into a limestone cliff face. The path of an ancient underground stream, this tortuous subterranean passage is the cave of Combarelles 1. At its entrance, a guide unlocks an ancient iron grille and swings it open. Beyond, a low, narrow winding tunnel disappears into the gloom, barely high enough to stand up in or wide enough to squeeze past your neighbor. Weaving and ducking, you proceed a hundred and fifty yards along this somber passage, wondering why on Earth you have come to this cramped, forbidding place, dark despite being lit at intervals by electric lights. Suddenly the guide stops, and all your questions vanish. Under the oblique light of filtered lamps, the walls of the cave suddenly come alive with engravings, some of them almost obscured by a calcite coating deposited on the cave walls over the millennia. Horses, mammoths, reindeer, bison, mountain goats, lions, and a host of other mammals cascade in image along the cave walls over a distance of almost a hundred yards, over three hundred depictions in all. Delicately executed and meticulously observed, these varied and overlapping images were made by people of the late Ice Age, perhaps thirteen thousand years ago.

You are mesmerized, not simply by the subtlety of these marvelous engravings—done at a time when the landscape around Combarelles, now oak forest, was open steppe roamed by mammoths, woolly rhinoceroses, and cave lions—but by their sheer ancientness. For this is not in any sense crude art; it is art as refined in its own way—and certainly as powerful—as anything achieved since. Any preconceptions you may have had of the "primitiveness" of "cavemen" are instantly dispelled. When you finally tear yourself away from these superb relics of an unimaginably remote human past and turn to grope your way back to the cave entrance, you notice something else. Halfway up the cave wall, the nature of its surface changes. The bottom of the cave has been dug out in this century, to make it easier for you to visit. When the Ice Age artists entered the cave, it was in places less than two feet high. These early people had had to squirm their way in on their bellies, squeezing between the cave's ceiling and floor, doubtless gasping for breath in the oxygen-poor air. Once a distant point had been chosen for decoration after the best part of an hour's uncomfortable crawling, the artists barely had room to move their arms, let alone the rest of their bodies. At the same time, they must have carried with them into the cave not only the flint tools that they used for engraving, but unwieldy and unreliable sources of light. Illumination, as we know from other sites of similar age, was supplied by lamps consisting mostly of hollowed-out slabs of rock (occasionally quite elaborately shaped and decorated), in which lumps of animal fat burned through juniper wicks. Though feeble and easily extinguished, these guttering lamps must have added a remarkable effect to the engravings as the artists experienced them. For in flickering light such as they produced, the images seem to come alive, bouncing along the cave walls; and the artists rendered the animals in suitably active poses.

Leaving the cave, you are consumed with the question "Why?" Why wriggle and struggle along a constricted, choking, dark, uncomfortable, and potentially dangerous passage that dead-ends deep in the rock with barely room to turn around? Why create art that could only be revisited with the greatest of difficulty? Why virtually ignore the outer part of the cave, executing your art only in its far interior recesses? Why engrave image over image, and intersperse those lifelike renderings with geometric signs and a profusion of obscure and apparently superfluous markings? And, quite simply, why the art at all?

Frankly, we will never certainly know the answer to any of these questions, although it's fun to speculate and we'll do so in the next chapter. This remarkable art is the miraculously preserved symbolic expression of the yearnings and values of a culture that has long disappeared, leaving us only these indirect and shadowy reflections of the doubtless rich body of myth, belief, and tradition that they embodied. But whatever obscure compulsion it was that propelled the artists into the dank recesses of Combarelles and many other caves in France and Spain all those many millennia ago, we can instinctively recognize it as something profoundly human. Not only is it humans, uniquely, who create art, but it is only we who indulge in behaviors as mysterious and unfathomable as this.

We human beings are indeed mysterious animals. We are linked to the living world, but we are sharply distinguished by our cognitive powers, and much of our behavior is conditioned by abstract and symbolic concerns. This isn't to say that we do not share various behaviors, proclivities, and physical structures with other kinds of animals: of course we do. Indeed, it is through such shared similarities that we know we form an integral part of nature, and it is by looking at the way in which they are distributed among the world's many species that we can recognize exactly where it is we fit into the great branching tree of life. Elaborating upon the structure of that tree is beyond the scope of this book; rather, I want to broach the question of human *uniqueness* by looking at what it is that sets us apart from our closest relatives in nature. These relatives come in two kinds. Among all living organisms, we are without question most closely related to the great apes, and part of this book will be devoted to looking at the size of the cognitive gulf between them and us—for we can only directly assess cognitive qualities in living animals.

In the larger scheme of things, however, the apes and we are not particularly closely related. We shared an ancestor with one (or more) of them that lived at least six to seven million years ago; and much has happened since, in their lineages as well as in ours. The archives of those happenings are our fossil and archaeological records, which contain information on the physical structure and behaviors of at least a dozen human and prehuman species, some very closely related to us, others less so. What we can glean of cognition from consideration of such things as brain size or the structure of archaeological sites is, of course, entirely inferential. We cannot directly know how human ancestors or extinct relatives behaved, or thought. But, by combining an understanding of our closest living relatives with the record left behind by our extinct ones, we can hope to gain some insight on what it is that makes us special and on how we acquired our uniquenesses. What's more, examining the underlying processes by which we became what we are helps us to understand not only our past, but the prospects for our future. All these are my subjects here.

CHAPTER *1*

. . . . . . . . . . . . . . . . . . . . . . . . . .

# *The Creative Explosion*

Human beings, in all their uniqueness, are the result of a long evolutionary process; and it is this which will be the central subject of this book. But since we started in Ice Age France, let's begin our evolutionary journey at the near end, so to speak, with a look at the astonishing record left by the Europeans of the late Ice Age. For these people provide us with the earliest good record of the unique human capacity, fully formed: evidence for what the science writer John Pfeiffer has called "the creative explosion." Not that this was an indigenous development; Europe was, until about forty thousand years (40 kyr) ago, inhabited only by the Neanderthals: a distinctive and now-extinct group of humans belonging to the species *Homo neanderthalensis.* The Neanderthals, whom we'll meet again in chapter 5, were complex beings and talented users of the landscape they lived in: a far cry, indeed, from the brutish image with which generations of cartoonists have endowed them. But they left no evidence of the creative, innovative spark that is so conspicuous a characteristic of our own kind; and they were quite rapidly displaced by the first European *Homo sapiens,* who arrived at that time fully equipped with modern behaviors.

These new Europeans are often known as Cro-Magnons, from the site in western France whence their fossil remains were first described. Exactly where the first Cro-Magnons arrived *from* is still not very clear (we'll return to this in chapters 5 and 6); but there's no doubt that they were *us*. Physically they were indistinguishable from living *Homo sapiens;* and, in its richness and complexity, the surviving material evidence of their lives indicates unequivocally that they were our intellectual equals.

These early Europeans were hunters and gatherers: people who lived off the resources available on the landscape. They arrived in their new land at a time when the climate was cooling considerably and the northern polar ice cap was building toward its maximum southward extent. By about 18 kyr ago, the edge of the northern ice sheet had crept south to the latitude of northern Germany and southern England, and beyond it stretched vast expanses of largely treeless landscape over which large-bodied grazing mammals moved in vast numbers. Cold times were thus not necessarily hard times for the first anatomically modern Europeans—although during certain periods, at least, the skeletons of Cro-Magnons often bear witness to difficult lives. For skilled hunters with all the cognitive powers of modern humans, the abundant fauna of the open steppic landscape was an incomparable resource to be exploited, sometimes with relatively little effort. And Cro-Magnon sites testify that these people took full advantage of what was available to them. The variety of animal bones left behind at places where Cro-Magnons camped vastly surpasses anything found previously: bird and fish bones, for example, show up virtually for the first time. This is not to say that Neanderthals, for example, never caught fish; bears do, after all. But if they did, they ate them on the spot, whereas the Cro-Magnons took them back to camp to be shared by all in typically modern human fashion. Dramatic evidence for such sharing comes from one locality in France, where archaeologists have identified the remains of a single animal distributed between three different campfire sites separated by hundreds of feet and presumably occupied by different families.

The Cro-Magnons also had an unprecedented knowledge of the habits of their prey. We see this not only in the wide range of animals they consumed, but in the placement of their camps and in their art. Many sites lie close to places at which herds of such mammals as reindeer would have had to ford streams, at which time they would have been particularly vulnerable to ambush hunters; and vast accumulations of animal bones, sometimes showing evidence of cooking, have been found in association with stone tools at the ends of blind valleys into which the victims must have been stampeded, or at the bases of cliffs over which they must have been chased. We know for certain that the Cro-Magnons carefully monitored their prey over the seasons of the year: animal depictions sometimes show bison in summer molting pelage, stags baying in the autumn rut, woolly rhinoceroses displaying the skin fold that was visible only in summer, or salmon with the curious spur on the lower jaw that males develop in the spawning season. Indeed, we know things about the anatomy of now-extinct animals that we could only know through the Cro-Magnons' art. For while soft-tissue features do not normally survive in the fossil record, they do so on cave walls and on small engraved slabs. We know only from the record left us by the Cro-Magnons, for example, that the extinct rhinoceroses of Ice Age Europe were adorned with shaggy coats, and that the extraordinary *Megaloceros giganteus*, a deer with vast antlers whose most recent bones date from 10,600 years ago, bore a dramatic and darkly colored hump behind the shoulders. The sole exception to the nonpreservation of soft structures, the frozen carcasses of extinct woolly mammoths found in the wastes of Siberia, serves also to emphasize the perceptiveness of Ice Age artists. For their peculiar features are exquisitely preserved on the cave walls, right down to their remarkable split-tipped trunks.

There's much more in the long record of Cro-Magnon life between about 40 and 10 kyr ago that is totally unprecedented in the record now available to us. Campsites were much more varied in size and complexity than anything earlier and if in sheltered spots, were usually placed to catch the warmth of the morning sun. Elaborate shelters were rigged up at open sites and were often much more complex than bare necessity demanded. The most remarkable such structures are known from localities on the central European plain about 15 kyr old. At the Ukrainian site of Mezhirich, the remains of four huts are known that were covered with complex arrangements of mammoth bones: tons of them. The deliberate and individualistic way in which the bones were chosen and disposed on each hut has led them to be dubbed the "earliest architecture." One hut is distinguished by a careful herringbone pattern of mammoth lower jaws; another by a palisade-like ring of long bones placed on end. At this same site, and others, it appears that the inhabitants dug pits in the permafrost: natural freezers in which meat was stored. This innovation may have allowed a semisedentary existence, the inhabitants living off their reserves of meat even when the migratory herds on which they depended had moved away. Uses of fire, which had been mastered in a rudimentary way even before the Neanderthals' time, became much more imaginative. In addition to the lamps used to light cave interiors, for example, elaborate hearths were constructed in a variety of styles, and it appears that hot stones were used to heat water in skin-lined pits. As long as 26 kyr ago, Cro-Magnons in what is now the Czech Republic were even baking clay statuettes (and maybe, for obscure ritual purposes, deliberately fracturing them) in kilns that heated to eight hundred degrees Fahrenheit.

Stone tools had been made for two and a half million years

by the time the Cro-Magnons came on the scene, but the Upper Paleolithic stone implements brought by these people to Europe show unsurpassed technological skill. The basic technique involved shaping a large "core" of rock, preferably flint, into a cylindrical form from which numerous long, thin "blades" could be struck with a hammer that was normally made of wood, bone, or antler. The blades thus produced had long, sharp cutting edges and were modified into a variety of more specialized implements. Many of these were then hafted into wood or bone handles. This approach to stone toolmaking provided as much as ten times more cutting edge per pound of raw material than any technique ever used before; and routine hafting provided unprecedented versatility and effectiveness. For the first time, moreover, bone and antler were made into carefully crafted utensils. Expertly carved tapered bone points were made, and antlers were straightened to form spear-throwers, often elaborately shaped and decorated. These rodlike devices, still used by Eskimos and Australian Aborigines in historic times, have a hook at the back in which the base of the spear is placed, while the front is held by the hunter. Effectively increasing the arm length of the user, they allow spears to be hurled farther and more accurately than those simply launched from the hand.

By around 18 kyr ago or a little less, sophisticated fishing is indicated by barbed harpoons, sometimes with blood grooves to enhance their effectiveness, and by simpler devices that look like fish hooks. At about the same time, clay impressions reveal that vegetable fibers were being plaited into ropes. Tiny fine-eyed bone and antler needles were made as long as 26 kyr ago, announcing that carefully tailored clothing had arrived on the scene. This list of Cro-Magnon innovations could go on and on; for these people, already formidably equipped on their arrival in Europe, continued to add to their material and behavioral complexity with an amazing wealth of ingenuity and invention. Nothing like this appears in the record left by any earlier humans. Truly, a new kind of being was on Earth.

The Neanderthals had occasionally practiced burial of the dead, but among the Cro-Magnons we see for the first time evidence of regular and elaborate burial, with hints of ritual and belief in an afterlife. The most striking example of Cro-Magnon burial comes from the 28-kyr-old site of Sungir, in Russia, where two young individuals and a sixty-year-old male (no previous kind of human had ever survived to such an age) were interred with an astonishing material richness. Each of the deceased was dressed in clothing onto which more than three thousand ivory beads had been sewn; and experiments have shown that each bead had taken an hour to make. They also wore carved pendants, bracelets, and shell necklaces. The juveniles, buried head to head, were flanked by two mammoth tusks over two yards long. What's more, these tusks had been straightened, something that my colleague Randy White points out could only have been achieved by boiling them. But how? The imagination boggles, for this was clearly not a matter of dropping hot stones into a small skin-lined pit. Also found at Sungir were numerous bone tools and carved objects, including wheel-like forms and a small ivory horse decorated with a regular pattern of tiny holes. The elaborate interments at Sungir are only the most dramatic example of many; and taken together, these Cro-Magnon burials tell us a great deal about the people who carried them out.

First, in all human societies known to practice it, burial of the dead with grave goods (and the ritual invariably associated with placing such objects in the grave) indicates a belief in an afterlife: the goods are there because they will be useful to the deceased in the future. Grave goods need not necessarily be everyday items, although everything found at Sungir might have been, since personal adornment seems to be a basic human urge that was expressed by the Cro-Magnons to its fullest. But whether or not some of the Sungir artifacts were made specifically to be used in burial, what is certain is that the knowledge of inevitable death and spiritual awareness are closely linked, and in Cro-Magnon burial there is abundant inferential evidence for both. It is here that we have the most ancient incontrovertible evidence for the existence of religious experience.

Second, the sheer amount of effort put into the aesthetic productions found in the graves suggest that decoration, elaboration, and art were integral components of the lives and societies of the people who made them; they were no haphazard doodlings. Art was emphatically not an occasional or incidental occupation among these people; it was central to their experience of their environment and to the way they explained the world—and presumably also their position in it—to themselves.

Third, the societies concerned must have been running considerable economic surpluses to have allowed the disposal in this way of objects that were so valuable in terms of the time taken to make them. These people clearly didn't have to devote all their time to the basic business of making a living; they were efficient enough exploiters of their environment that leisure was available for symbolic pursuits of this kind. However, it's also fair to note that artistic production during the Ice Age was carried out in many environments, some of which were considerably more productive than others from the point of view of human hunters and gatherers. Once the practice of producing symbolic artifacts had become established, along with the ritual systems of which they formed part, it may well have been that artistic production became an integrated part of the economic system, viewed by Cro-Magnon societies as essential for maintaining their economic lives. When harder times arrived, as they must surely have done in the fluctuating climates of the Ice Age, these people may have seen their art and its associated ritual as something that was somehow necessary for their continued well-being: essential to their success in the hunt and in the other activities that sustained them from day to day.

Fourth, the fact that there is a considerable variety in the elaborateness and detail of Cro-Magnon burial (for the sheer opulence of Sungir is one exception, rather than the general rule) hints at a social stratification and division of labor in Cro-Magnon society. Richness of personal adornment in life often reflects social status, and this is in turn often mirrored by the objects taken to the grave. Some Cro-Magnons were buried with an extraordinary abundance of artifacts of various kinds; others were more simply interred. And while, given the erratic sampling of Cro-Magnon burials that we possess, some of this variation may simply have been due to difference in affluence of societies overall, there can be little doubt that part of it, at least, reflects a differing importance of individuals in society. Some of that difference in status may well have been inherited; for it is highly dubious that, for instance, the children of Sungir would have had the opportunity to make any significant mark on their society through their own accomplishments. What's more, it's hardly probable that these young people had made their richly adorned vestments themselves. It's much more likely that the sheer diversity of material production in their society was the result of the specialization of individuals in different activities. Those who ground the mammoth-tusk beads of Sungir, and who—by who knows what magic—straightened out those mammoth-tusk spears, may well have received far less elaborate interments when their own turn came to be buried.

Contrasting the Sungir site with other Ice Age burial localities also draws attention to the considerable local variation in mortuary practices that existed during Cro-Magnon times. In some places bodies were flexed in the grave; at Sungir they were stretched out. Some graves were covered with rock slabs; others weren't. The nature and abundance of grave goods varied from place to place. And on and on. Differences of such kinds in burial customs must, moreover, have reflected a wider cultural diversity. For example, in contrast to the relatively uniform material productions of earlier peoples, Cro-Magnon traditions of stone artifact making differed wildly from place to place. Sometimes, it seems, the people of each valley were busily developing their own particular ways of doing things, and it's even been suggested that hand in hand with this went linguistic diversification and the development of local dialects.

In this restless, innovative spirit we see our own modern selves mirrored, and the principal lesson to be learned from Cro-Magnon burials is this: That while we will never know exactly what rituals accompanied them and what exact sets of beliefs they embodied, these interments, taken overall, reflect not only the fundamental human urge to adorn and elaborate, but also the multifaceted subtlety and complexity of living human societies the world over.

Nonetheless, the material aspect of Cro-Magnon life that speaks to us most directly as human beings lies in the evidence these people left behind of art and symbolic representation. The astonishing art of the caves is well-known. But here our notion of art has to be used in its widest sense because some of the first Cro-Magnon sites have yielded evidence for music and notation as well. The earliest Cro-Magnon culture identified in Europe is known as the Aurignacian; and some of the oldest Aurignacian localities, dating from well over 30 kyr ago, have produced musical instruments: multiholed bone flutes capable of producing a remarkable complexity of sound. Later sites have also yielded what may have been percussion instruments; and at one locality, a series of enormous flint blades found laid parallel on the ground may have been the remains of a "lithiphone": the Stone Age equivalent of a xylophone. The earliest Aurignacian has also yielded bone and stone plaques bearing extremely complex markings; one 32-kyr-old plaque from the French site of Abri Blanchard has been identified as a lunar calendar, and many objects incised with regular patterns of marks have been interpreted as hunting tallies or other forms of record keeping. That's all as may be; we will never be certain exactly what particular abstract symbols meant to their long-departed creators, however evident it is that they were intentionally made. This is true even of what appears to our eyes to be representational art; for while we may readily recognize as such the animals depicted on cave walls and stone and bone plaques, to their makers they may well have been symbolically equivalent to the more obscure geometrical or superficially random-appearing markings that baffle us from the start. What is obvious, however, is that here we have evidence of highly complex symbolic systems.

Still, as I've said, among all the legacies left to us by the Cro-Magnons, it is what we instinctively feel to be "art" that most readily captures our imaginations. Art as such, of course, is a concept invented by Western civilization. The universal human urge to decorate aside, what we recognize as "art" produced by other cultures in the modern world tends to be quite distinct in significance from the aesthetic notions we attach to art in our own culture; and the same was evidently true of the Upper Paleolithic. Searching for the "meaning" of Ice Age art in the absence of the living society that produced it is thus likely to be unproductive. What we can do, however, is to develop a chronology for this art and to look for regularities in it that may help us to understand its structure. Chronology is especially important here because Ice Age art was not the outpouring of a single culture. Rather, it spanned a period of over twenty thousand years within which several cultures, as recognized by their technological traditions—archaeologists' normal touchstone—came and went. Thus, remarkably, the earliest European Ice Age art was over twice as remote in time from its latest expressions as the latter are from us. Yet working out the chronology of Ice Age art is turning out to be trickier than once thought.

Until quite recently it was generally believed, for example, that the practice of making paintings in deep caves was a relatively late development, getting under way—slowly—only about 25 to 24 kyr ago, 10 kyr later than the first three-dimensional sculptures. Of course, paintings on cave walls have always posed a problem of dating because they are free of any archaeological context. In contrast, "portable" art, images carved and engraved on small pieces of bone or rock, is always found in the layers of habitation detritus left behind by early people. Where a given site was inhabited, consistently or sporadically, over an extended period, later strata of this kind accumulated on top of the earlier ones, providing a "layer cake" sequence that maps cultural changes over time. In habitation layers, artworks are associated with the stone tools upon which Paleolithic ("Old Stone Age") chronologies have traditionally been based via the mapping of technological change. Further, cultural strata sometimes also contain organic objects that are directly datable by the carbon-14 method (good up to about 40 kyr ago, hence ideal for Cro-Magnon times). By combining studies of style in portable art with their archaeological contexts, students of Ice Age art have been able to develop a chronology of art styles over the various periods of the Upper Paleolithic: the final period of the Old Stone Age during which the Cro-Magnons flourished.

Even this laborious procedure is only rough-and-ready, however, especially since many of the most striking and important pieces of portable art were excavated in the early days of archaeology, when relatively little attention was paid to context. What's more, "styles" are often hard to recognize. But, added to consideration of the way in which images executed in different manners are superimposed on cave walls, it has been this stylistic chronology that has largely governed our ideas of the sequence in which the hundreds of works of cave art now known were created. Excitingly, though, direct dating of cave art has very recently become possible (by new radiocarbon techniques), although only in those few cases where the artists used organic materials such as charcoal in executing their works. And the little that has so far been learned from these new approaches has rocked our notions of the chronology of Ice Age art to their foundations.

For it has turned out that, far from being a relatively late development among Ice Age cultures, deep cave art was quite an early innovation. Thus, dating of the paintings in the newly discovered cave of Chauvet, in south-central France, has shown that at least some of the 300-plus animal images that cascade across the cave walls there date back to over 30 kyr ago. This contrasts dramatically with more traditional estimates of their age, made shortly after their discovery, of maybe 18 kyr. Nobody had imagined that wall art of this sophistication could be of such remarkable antiquity; but similarly early ages have since been derived from other sites, and it's clear that the chronology of Ice Age art is due for radical reappraisal. It's still unquestioned, however, that the golden age of Ice Age art—portable and wall alike—occurred during what is called the Magdalenian period, which began at about 18 kyr ago—interestingly, just as the last glaciation was reaching the maximum of its intensity.

What does all this tell us about the Cro-Magnons? First, it dramatically bolsters the conclusion that the first modern people arrived in Europe equipped with all of the cognitive skills that we possess today. Second, it underlines the tendency toward innovation and cultural diversification that is so fundamental a characteristic of *Homo sapiens*—and so foreign to all earlier human species. Some investigators have believed that—using the old chronology—they could trace a single strand of stylistic evolution all the way through Ice Age art, from the first productions of the Aurignacian at about 34 kyr ago, right up through the end of the Magdalenian at about 10 kyr ago. Such continuity was always inherently improbable over such a vast span of time, and it is now clear that this was not the case. At the same time, the pattern of sporadic technological innovation over the Upper Paleolithic still appears to apply: bone spear points appear at over 34 kyr ago, for instance, while bone needles do not show up until about 26 kyr and barbed harpoons not until about 18 kyr ago.

A lack of continuity makes it easier to understand—though no less astonishing—how the very earliest art we know of includes some of the finest creations of all time. At the German site of Vogelherd, dating to the earliest Aurignacian of the region, over 32 kyr ago, a series of small animal figures testifies to the highest standards of the carver's art. The most striking of these pieces is a horse, barely two inches long, made from mammoth ivory. Polished from long contact with someone's skin, this tiny figure, probably worn originally as a pendant, is one of the most elegant images ever carved. It does not closely resemble the chunky, stocky horses that roamed the European steppes at the time; rather, in its flowing, sinuous lines, it evokes the graceful essence of the horse. No crude representation this; its maker possessed both skill and imagination to rival any later artist. Symbolism of other kinds also made an early debut, as we've seen—and not just in the form of notation on plaques. Perhaps even earlier than Vogelherd is a foot-tall carving, from the nearby cave of Hohlenstein-Stadel, of a standing human with a lion's head. This is less finely executed than the Vogelherd horse; but what better reflection could we wish for of the exercise of the human imagination? This is surely a mythic figure, an embodiment of a complex mixture of myth, observation, and belief that explored and explained the place of humanity in the larger world.

As I've suggested, though, not all innovations appeared together. For the late Ice Age was remarkable for the cultural diversity it spawned, in time as well as in space. Following the Aurignacian, we find the Gravettian culture, most noted for its production of the female statuettes and engravings misleadingly known as "Venus" figures. These typically show women with heads lacking individual features, swollen midbodies with large breasts and buttocks, and small limbs. Some "Venuses," however, are more linear and modestly proportioned. These figures have traditionally been interpreted as fertility symbols, but in view of the fact that fertility is rarely an issue among hunting-gathering peoples, this is hardly a convincing explanation. Nonetheless, the fact that they are found over a huge swath of Europe over a long time period (between about 28 and 22 kyr in western Europe, until much later in the east) strongly suggests that these images were embedded in a powerful and durable cultural tradition. Still, local variants are clearly evident within this tradition. Perhaps most remarkably, at the Czech site of Dolní Věstonice, female and animal figurines were baked and apparently deliberately fractured in kilns during what were probably homesite rituals of some sort. This kind of production is known from nowhere else; and, indeed, the notion of baking clay subsequently lay fallow for as many as 150 centuries, until pottery was introduced in the New Stone Age, this time in the service of utilitarian purposes.

As I've already briefly noted, though, the most remarkable artistic expression of the Upper Paleolithic was that of the Magdalenian culture, which began at about 18 kyr ago. The Magdalenians had at their disposal the widest panoply of technological gadgetry, and it was during their tenure that the finest cave art was created, including that of the spectacular sites of Lascaux, Altamira, and Niaux, and that the most remarkable outpourings of portable art were made. Which brings us back to the question posed earlier, as we left Combarelles: Why cave art?

Let's start with the fact that the cave art sites themselves are very variable—just as is the quality and the quantity of the art we find inside them. Some are tiny, cramped spaces, which only a few individuals could enter at any one time. Some are huge caverns, with the potential to contain a large number of spectators. Some are decorated in places that are easy to reach; in others, art is found only in the remotest and most difficult recesses. At one cave in France, you have to paddle up an underground stream, then walk, wriggle, and crawl for two hours before finally, a mile underground, reaching the terminal chamber. Here, maybe 14 kyr ago, Magdalenians sculpted, out of clay taken from the cave floor, a pair—male and female—of magnificent bison, each almost two feet long.

Today, entering such caves is an alien experience, even with the benefit of electric light, and you feel that the Magdalenians must have been extraordinarily courageous to have undertaken such subterranean journeys by the faint and vulnerable light of fat lamps. But at one point in the bison cave, far underground, there is a muddy ledge that bears the footprints of a child, no more than six years old, who had pranced down the ledge, digging in his or her toes—you can still see the imprints—just before the drop-off to the cave floor a foot below. At another site you can see where, a thousand yards underground, an adult must have taken the hand of a child and traced its finger across the soft surface of the cave wall. And at yet another, you can see where an adult had held a child's hand against the cave wall and blown pigment over it, leaving on the wall a negative imprint of the tiny fingers and palm.

What are we to make of observations such as these? If Ice Age adults had taken children into these dark caves, in the recesses of which it is easy even for modern explorers to lose their way, it is hard to conclude that they found such places as frightening and forbidding as most of us do. Indeed, it seems likely that the people of the Upper Paleolithic felt quite comfortable in them. But it also seems clear that these were nonetheless *special* places to the Cro-Magnons. For example, there is very little evidence that anyone ever lived deep inside Ice Age caves or carried out mundane activities there. People often lived in cave entrances, where shelter was combined with light and space; but they appear normally to have penetrated deep inside the pitch-black cave interiors only for special purposes that were presumably associated with the art they left there—and repeated visits to the art are frequently attested to by overengraving and alteration of the images: the art was "used" over and over again.

The Cro-Magnons did not, however, exclusively create wall art deep in caves—though this is where for the most part it has been preserved. It is quite probable that they often decorated their living spaces, too, although the evidence for this practice in places exposed to the elements has, with only a few exceptions, disappeared. One of these exceptions is quite extraordinary, though. At the rock shelter (a space beneath an overhanging limestone ledge) of Cap Blanc, in western France, an extraordinary frieze has been preserved. Here, over a length of fifteen yards, fourteen animals (mostly horses), some of them almost full size, tumble in side view across the shelter wall in deeply incised relief and with a fine sense of composition and even of perspective, as reflected in their arrangement on the wall. Traces of pigment show that these images were originally painted reddish brown, using ocher that occurred naturally in the earth around; and it is quite possible that during Magdalenian times, the walls of the limestone valleys of this region were in many places a blaze of color.

It also seems likely, however, that cave painting had a different significance to the Magdalenians than the decoration of their living sites had. One pointer to this—quite apart from the probability that cave art was quite rarely visited whereas portable art was evidently an integral part of Cro-Magnons' daily lives—lies in the choice of animals represented in caves, which does not reflect the abundance of species on the landscape. Horses, wild cattle, mammoths, and bison are there in the caves in their hundreds; but reindeer (a major prey species) are rare, and roe deer, weasels, and tortoises almost absent. What's more, the representation of animal species in cave art is very different from that in the portable art found at living sites, where among thousands of images there are many more reindeer and far fewer mammoths (and no weasels at all), and the range of species represented is also much wider. In the portable art, then, we find a much truer representation both of the overall fauna and of prey species. Caution is in order, though, because in those parts of Europe in which both art forms were practiced, there are rather few sites that are rich in both. It's also interesting to note that while human images are commoner in Ice Age art than generally thought, they are almost invariably significantly less realistic than those of other animals—especially on the cave walls.

In most caves, decoration is sporadic and limited to a few images at best. In some, however, the sheer abundance and exuberance of image is no less than overwhelming. This is particularly true of the smallish cave of Lascaux, in western France, decorated about 17 kyr ago and probably more or less at one time. I can think of no more powerful experience I've ever had than my own first entry into this extraordinary place, where the first chamber you enter (and there are others, equally impressive) is a riot of large multicolored animal images, dominated by horses and huge wild cattle, and including one of a possibly mythical beast. In observation, drama, and sheer skill of representation (on the most difficult of surfaces), Lascaux remains unsurpassed to this day; these are no simple renderings, and the artists drew on subtle tricks of perspective to make their images even more compelling.

Typically, the animals at Lascaux are complemented by a host of regular geometric signs; and although we will never know what these signs meant to those who drew them, there is clearly a lot more going on here than meets the modern eye. But one aspect of the creations of Lascaux is a rarity in Ice Age art, in that what we can readily recognize as a scenario of some kind is depicted there. At the bottom of a pit in the cave, there exists a complex scene, in which a (typically schematic and in this case bird-headed and ithyphallic) man falls before a huge bison that appears to have been disemboweled by a spear piercing its flank. Behind the human, a woolly rhinoceros retreats, and below him is a pole with a bird perched atop. If anything in Ice Age art represents a mythic scene, this one does; and, interestingly, at the cave of Villars, some thirty miles away, a similar if simplified drama appears to be enacted. How can we fail to speculate on the meaning of a scenario such as this? Certainly, the great French archaeologist François Bordes couldn't resist. "Once upon a time," he surmised, "a hunter who belonged to the bird clan was killed by a bison. One of his companions, a member of the rhinoceros clan, entered the cave and drew the scene of his friend's death—and of his revenge. The bison is pierced by weapons and is disemboweled, possibly by the horns of the rhinoceros." And while—as Bordes himself knew very well—his story is almost certainly wrong, this scene, whatever its meaning, does evoke a peculiarly human consciousness. As you leave Lascaux, you are overawed by the magnificence of what these remote ancestors wrought all those many millennia ago; and, however different their lives, language, and beliefs may have been from yours and mine, you also feel an extraordinarily close identity with them.

The intensely social nature of the art of Lascaux is emphasized by the certainty that it was the work of a team. There are clear indications in the cave that scaffolding was rigged up to allow decoration of the walls well above a height that could be reached by an artist standing on the cave floor; and a clay impression was found of a rope that may have bound the scaffolding together. Reindeer bones found in the cave may have been the remains of meals taken by the crew; and dozens of stone lamps scattered around on the cave floor (one of them a work of art in itself) attest to an elaborate approach to lighting the dark cave interior. These details alone suggest that the art of Lascaux was the work of specialists; but the best evidence for this comes from the sheer excellence of the images themselves. Lascaux was indeed a "sanctuary" of some kind, of great importance to the society that produced it—though a sanctuary of exactly what kind, reflecting what set of beliefs, we'll never be certain.

Early students of cave art tended to the view that the artistic creations of Ice Age people represented the exercise of "hunting magic," a notion also reflected in the more recent Bordes interpretation of the mythic scene. In such constructs, the art embodies the importance of the animals represented in the economic lives of the people who made it. Through "sympathetic magic," the depiction of the animals would assist their fecundity and thereby enhance the chances for survival or prosperity of the humans who depended on them. Moreover, many depictions of animals in Ice Age art are overlain by apparently extraneous lines, which certainly look as if they might represent spears; and, in a few places, human images are also "pierced" in this way. Such symbolic "wounding" might be directed at

*Monochrome rendering of a badly faded polychrome wall painting in the cave of Font de Gaume, France. A female reindeer kneels before a magnificently antlered male, who leans forward and delicately licks her forehead. Probably painted about 14 kyr ago.*

enhancing success in the hunt—or, ominously, in warfare. What's more, given their inaccessible location deep in caves, it was believed that the images remained the property of an elite "sorcerer" class, who held the keys to the economic success of the group and maintained a status separate from that of the others who supported them. All this hinted darkly at ritual, something approaching a religion based on assuring economic security. Success in the hunt, based as it certainly was on the sheer skill and effort of the hunters, would have appeared in these peoples' retrospect as proof of the efficacity of such ritual.

Such interpretations are no longer in vogue. It was pretty soon pointed out, for example, that the animals depicted in caves were not necessarily those most commonly hunted. Mammoths, bears, and woolly rhinoceroses would not have been the first choices for hunters who wished to survive for long, while reindeer, the staple of the Upper Paleolithic diet, are rarely seen on cave walls. Moreover, the most famous image of reindeer in Ice Age cave art is one that we would hardly associate with killing. At the French cave of Font de Gaume, there is a badly faded but intensely moving image of two reindeer. On the right is a female, bending down, forelegs flexed. Before her stands a male with magnificent antlers, leaning forward and gently licking her brow. The scene is beautifully observed and full of tenderness; and I, for one, am incapable of viewing it in any violent context. Some other factor than the hunt must have been in the mind of the creator of this masterpiece.

Still viewing caves as closed sanctuaries, researchers thus began to look at how the decoration of these spaces was structured. It was found—as a general rule, with plenty of exceptions—that the "central" decorated areas of caves contained depictions of horses, bison, and wild cattle; that these tended to be surrounded by images of mammoths, deer, and mountain goats; and that bears, lions, and rhinoceroses tended to be found in the deepest cave recesses. It was also found that, at least regionally, animal images tended to be paired: horses with bison or their close relatives wild cattle, for example. It was suggested that this pairing reflected the opposition of the sexes, and that the gender of a member of a particular species was independent of the sex that species represented. This rather Freudian interpretation hinted at an elaborate gender-based system of symbolism, the origin and consequences of which remained unclear. Accumulating knowledge of the caves has rendered such pairings less conspicuous, though, and this notion has recently had a rather rough ride. The new knowledge on which it was based has, however, reinforced the perception of the decorated caves as being special, consciously structured places, of profound and particular significance to those who entered them. Leaving such precincts, inevitably filled with deep if obscure emotion and wonder, no modern visitor could ever readily disagree.

But, especially given the "private" nature of the usage of some of the caves, with children of very tender age being conducted within and apparently being consciously encouraged to imitate the adults, we will never be sure exactly what the caves' significance was.

Still, many caves were decorated at intervals over a long period of time, and it seems possible that later artists respected the work of earlier ones. At the French cave of La Mouthe, for example, the existence of earlier artwork closer to the entrance appears to have "forced" subsequent artists ever deeper into the cave, thereby creating a largely illusory symbolic "structure" if we consider only the species represented. What this also emphasizes, however, is the durability of the heritage of Ice Age art. For over 20 kyr, the early people of the Franco-Cantabrian area continued at least sporadically to produce art—sometimes great art—on cave walls. It is inconceivable simply on grounds of time, let alone of the abundant evidence of cultural change in the Upper Paleolithic and of what we know of modern humans, that this art was the product of a single unbroken tradition; and, indeed, it is becoming evident on both intrinsic and extrinsic grounds that it was not. What was it, then, that lay behind the production of art in this particular place (for cave art is far more restricted in its geographic distribution than portable art), over this long period?

One intriguing suggestion is that the area in which Ice Age cave art is found provided, during the last Ice Age, one of the richest habitats in which human hunters and gatherers have ever lived. The late Ice Age inhabitants of southern France and northern Spain lived in an area greatly favored by geography. This relatively sheltered limestone region boasted a huge range of habitats, from the hilly crags, where ibex abounded, right down to the valley floors, where the rivers teemed seasonally with migrating salmon. Perhaps never before had fully modern humans ever lived in such a productive and varied environment; and if so, the visual stimuli to artistic production had never been so great. The art of the Cro-Magnons reflected the richness of their environment and doubtless also the ways in which they explained to themselves their place in it. And while it is improbable that this supplies a full explanation for the ubiquity, quantity, and long duration of Cro-Magnon art, the nature of these peoples' environment was undoubtedly a factor.

The acuteness and abundance of reference to that environment in Cro-Magnon art almost certainly indicates that the Cro-Magnons believed themselves to belong to the ecosystems surrounding them; that they felt themselves to be integrated into them. Interestingly, the French archaeologist Jean Clottes has recently noticed that the frequency of "dangerous" animals declines steadily over the long period of Ice Age art. In the Aurignacian, for example, images of such animals as lions, rhinoceroses, mammoths, and bears account for over 50 percent of the different kinds of animals represented. By the time the Magdalenian came around, that figure had dropped to around 5 percent overall (though there's a lot of variation between sites). This difference must surely represent a change in the way that the Cro-Magnons were viewing the world about them, and probably also in the way they were interacting with it. Perhaps the Magdalenians had simply evolved better ways of coping with predators, rendering them less threatening than they had been to their remote forebears, and thus less powerful images of external danger. Certainly, they had a more complex technological panoply for dealing with them. Or perhaps there had been a change in the ways in which these early people saw their own relationship to the external world and in the imagery they used to express this. We'll return at the end of this book to the

Cro-Magnons'—and our own—view of humanity's place in nature.

Whatever the case, the Cro-Magnons have bequeathed us dramatic evidence of the integral importance of symbol in their lives. Careful analysis of the abstract signs they painted on cave walls and engraved on plaques, as well as a better understanding of their "art," is beginning to reveal that we are not looking here at a single unitary body of symbol and belief. Rather, there were dozens of symbolic systems current during the last Ice Age, just in the geographically restricted region of western Europe. This typically modern human cultural diversity—and, indeed, the pervasiveness of symbolic behavior itself—stands in dramatic contrast to the relative monotony of human evolution throughout the five million years that preceded it. For prior to the Cro-Magnons, innovation was as far as we can tell extremely sporadic at best. New techniques are beginning to reveal facets and tempos in the evolution of Ice Age art and society that were undreamed of earlier, bringing this field of science into an unprecedented era of ferment. We still cannot see where this ferment will end; but it is already amply evident that with the arrival on Earth of modern *Homo sapiens*, a truly new kind of entity had emerged: one whose potential we are still exploring and enlarging today.

As I have hinted, there are suggestions of early art and symbolism elsewhere than in Europe, and doubtless in the late Ice Age, similar developments were taking place throughout the Old World. But the record left us by the Cro-Magnons is so far unparalleled. We will look in chapters 5 and 6 at other evidence for the rise and spread of the unique human capacity; but for the moment, at least, it remains true that it is against the well-documented behavior and achievements of the Upper

Paleolithic Europeans that those of earlier humans have to be judged. Later in this book, we will take up the long record of human evolutionary history, and as we do so, it will be useful to bear in mind the magnitude of the Cro-Magnons' cultural achievements.

CHAPTER 2

# *The Brain and Intelligence: Humans and Apes*

I remarked in the last chapter that with the appearance on Earth of modern *Homo sapiens*, a totally new kind of being had arrived on the scene. But just how special are we? Exactly what are the attributes that give us our acute sense of being so *different* from all other living species? How and why were these attributes acquired in the course of our evolutionary past? And, a particularly lively issue today, can we usefully seek evolutionary explanations for our complex and often unfathomable behaviors? We'll pursue all of these questions in the course of this book, but let's start at the beginning, with an assessment of our uniqueness among living organisms.

When the great apes were first discovered by literate peoples, back in classical times and even earlier, the intuitive reaction was to recognize in them a variant of humankind, as witness the first record of any ape, made by the renowned Carthaginian navigator Hanno. Exploring along the coast of West Africa in the fifth century B.C., Hanno encountered "savage people . . . whose bodies were hairy and whom our interpreters called Gorillas." Hanno knew humans and he knew monkeys; and, reading his account, you are left in no doubt with which group he felt his gorillas (which might in fact have been chimpanzees) belonged. But familiarity, it seems, eventually bred contempt; and by the time the Victorian era came around, the apes' image had become one of benighted far-from-humanness, zoo visitors seeing them at best as ludicrous caricatures of themselves. Today, though, we know a great deal more about the apes than our Victorian forefathers did; and what we have learned shows that they are considerably more complex beings than their public image implies.

Unlike Hanno, we now know why the apes resemble us so uncannily: in the entire living world they are our closest relatives, connected to us by recency of common ancestry. In a broad sense, we ourselves are apes, because humans and apes together are descended from something we can more easily call an ape than a human; but this ancestor was very evidently not like any of the apes living today, which possess a variety of anatomical specializations that the ancestor lacked. Nonetheless, among all living things, the great apes provide us with the best yardstick for assessing our own uniqueness. Of course, the apes are not our closest *known* relatives, for numerous fossil humans, whom we'll meet later, share more recent ancestry with us. The apes are, however, the closest relatives we can experience as living, breathing entities and whose behaviors we can observe directly. This ability to observe the apes in action gives us a direct prospect on cognition that we cannot hope to gain from the indirect evidence that the archaeological record yields about closer relatives that now have the misfortune to be extinct. Recent studies have shown that we share more of our attributes with the great apes than we have liked to imagine; but understanding the differences between them and us establishes the essential point of departure for interpreting the human fossil record.

## *What Is Intelligence?*

We differ from the great apes in the structure of numerous bodily systems; but then, this is the general nature of biological diversity. When we think about what differentiates us from these relatives, it's not our shorter faces and smaller canine teeth that spring most immediately to mind, or even the fact that we have difficulty lighting matches using our toes while they wouldn't—if they had matches. What truly makes us different from other living creatures is exactly what we *feel* makes us different: our unique cognitive capacities. Quite (and deceptively) simply, we are more intelligent than other species are. And not simply *more* intelligent—numerous species are more intelligent than others—but *differently* intelligent, in a manner that allows us not only to view ourselves, but also to manipulate the environment around us, in a qualitatively unique way.

Yet what do we mean by intelligence? This elusive attribute has consistently frustrated the attempts of psychologists to characterize it to everyone's (even their own) satisfaction. The problem, of course, is that intelligence, whatever its essential constituent(s) may be, comes in many forms. You may be smart in areas in which I'm completely incompetent, and (possibly) vice versa. As a dying Nigel Greene said to David Niven in a movie I saw (very) late at night not long ago, "You may be a good doctor, but you're a bloody awful spy." What's more, while assessing intelligence among humans is difficult enough, as a huge literature and numerous controversies testify, the problems are vastly magnified when we try to evaluate intelligence in other species. You or I may find a given test very clever and/or interesting; but it is by no means obvious that a chimpanzee, say, should find the same test fascinating in the least. And if it doesn't, why should it even attempt to perform? Further, even if we make response virtually compulsory by offering an artificial reward, such as a prized food, there is no particular reason why the chimpanzee's response should necessarily be appropriate in our terms—those on which we dreamed up the test in the first place. We perceive our environment in ways that clearly differ from those in which the apes perceive theirs; and perception is inextricably entwined with the expression of intelligence. Caution is hence in order.

This is not to say that all artificial tests—even some that are irrelevant to the normal experience of their subjects—are necessarily uninformative about cleverness. Indeed, I shall be quoting some research of this kind that is, I think, extremely revealing. It does suggest, though, that if (as I think we can reasonably assume) the principal use of intelligence is to help the individual act as efficiently or responsively as possible within its physical, biotic, and social environments, we will learn most about this elusive quality by supplementing artificial manipulations with the observation of how flexibly and innovatively individuals cope with their natural habitats—broadly defined to include other members of their own species. Of course, in analyzing the behaviors of individuals belonging to nonhuman species, we run the risk of inappropriately ascribing to them motives and intentions that they do not actually have; but careful scrutiny of circumstances and alternatives will often acceptably reduce this risk.

By adopting this approach, we will also manage to sidestep the old debate about the "innate" and "learned" aspects of behavior. Many earlier attempts to compare intelligence between species sought to discriminate between innate and learned behaviors, aiming to single out the latter on the theory that while learning abilities must surely be linked to intelligence, innate behaviors—those that are firmly implanted in the genetic substrate—tell us nothing about the qualities of flexibility and innovativeness that are so clearly part of intelligence as we understand it in ourselves. The difficulty here is, though, that virtually no voluntary behaviors are devoid of some genetic substrate. In one sense, I suppose, each individual—of whatever species—does come into the world as a *tabula rasa*—a "clean slate"—hence has in some sense to learn its repeated (that is, virtually all) behaviors; but the nature of the inherited *tabula* itself varies among individuals and, on average, among species. Further, there are many levels at which learned behaviors may be acquired, and apparently similar behaviors may result from the actions of different mechanisms. Most importantly, though, intelligence and learning ability are clearly not synonymous. *Understanding* new (or even familiar) situations is different from simply responding to them, whether in a reflexive or a learned way. Moreover, developing novel responses to them (something yet different again) demands a distinct departure from previously learned behaviors—even though the association of past experiences with present circumstances may often be an essential ingredient in problem solving.

A large literature has by now accumulated on "intelligent" behaviors of great apes and other vertebrates, both in the wild and under experimental circumstances. Inevitably, the tendency is to recognize intelligent behaviors as those elicited by situations in which we humans would bring an "intelligent" approach to the solution of what we would define as "problems." This is fine, for where else could we start? Yet careful examination of most such instances shows that, the great apes perhaps apart, none of the subjects of such studies brings to those problems the understanding of cause and effect that governs the behavior of humans confronting similar situations. Let's look at some of this research.

## *Behavior and Intelligence*

Over the past few decades primatologists and cognitive psychologists have quite intensively studied higher primate behaviors that may resemble—and hence foreshadow—our own. The goal has generally been to come up with and to test ideas on *why* certain primates became intelligent; and this in turn has mostly boiled down to attempts to identify the evolutionary imperatives that *caused* an increase in intelligence. This aim is only possible at all, of course, if evolution is viewed purely as an unbroken, straight-line, gradually incremental affair, driven by inexorable natural selection. And as we'll see in the next chapter, evolution is in reality a rather more complicated business than this. Nonetheless, we have a lot to learn from comparing ourselves to the great apes.

Over the years two major contenders have emerged as potential stimuli for increased intelligence (or, as I would prefer to put it, as reasons for the success of more intelligent individuals or species). These are the social milieu and the external habitat. In the case of the latter, most analyses have focused on the means of finding food, although all primates—in fact all organisms—seem to have no difficulty doing so, at least in favorable conditions. Some species simply snuffle, swoop, or swim around, hoping to blunder into the next edible item that comes along; and they seem to have done perfectly well up to the present without "intelligent" seeking of food. It's true, however, that primates, most of them dwellers in the forest canopy, exist in a more complex set of ecological situations. Forests, most especially the tropical forests in which primates typically live, are not the same year-round, and resource trees are irregularly scattered within them. Much has therefore been made of the fact that this or that primate must have a "mental map" of its home range that charts the location of trees yielding food not only in space, but in time as the annual cycle progresses. Rare is the tree that produces fruit, say, or young leaves year-round; as far as a feeder on such resources is concerned, the same forest is two totally different places in May and December. And although this is a problem that faces all primates—indeed, all tree dwellers—to a greater or a lesser extent, it is possible that some primates have a more sophisticated appreciation of these variations than others. Indeed, there is a certain amount of evidence that this is so. Orangutans may be a case in point: they have been recorded to take extremely short routes, through vast tracts of monotonous rain forest, between single trees that were only transiently in fruit.

For anthropologists, of course, the aspect of food seeking that exerts the most fascination is hunting. Humans were once thought to be alone among primates in obtaining sustenance this way, but in recent years this perspective has changed dramatically. The hunting prowess of chimpanzees has by now been well publicized, but baboons and South American capuchin monkeys have also been observed to kill small mammals. What really singles out chimpanzees from the others, though, is the routine and apparently organized nature of many of their hunts (though as we learn more about baboons, we may find that chimpanzees are not alone). At two sites in Tanzania, for example, chimpanzees regularly catch and eat birds and more than a dozen mammal species, among which a particular favorite is colobus monkeys. Smaller-bodied species—or juveniles of larger ones—are preferred, but chimpanzees have been seen to take prey weighing not much short of fifty pounds. Sometimes a hunt is clearly opportunistic, as when an individual chimpanzee simply happens upon a suitable victim. At other times, though, a hunt can be a fully fledged cooperative enterprise, involving several individuals. Indeed, a "colobus hunt" will more often than not involve the capture of two to four unfortunate monkeys (usually juveniles) by what looks like an organized party of chimpanzees, one or more of which climb trees after the colobus, while the others spread out on the ground. The apparently well-rehearsed nature of this activity does not mean that chimpanzees wake up in the morning with the intention of hunting, as humans might do; but once a group of colobus has attracted the chimpanzees' attention, the apes do seem to assess the situation carefully. A hunt is rarely started where the forest canopy is continuous, and the monkeys thus have a good chance of fleeing successfully through the upper branches. Instead, predation is normally carried out where the colobus are feeding in one or a few isolated trees, from which it would be much more difficult for them to escape.

Male Tanzanian chimpanzees hunt more than females do, but female participation in group hunts is not unusual; and, indeed, all-female parties of chimpanzees have not infrequently been seen hunting. What's more, females take a higher proportion of the easier-to-catch prey, such as juvenile bushbucks and bushpigs, although males take the vast preponderance of more difficult victims, such as colobus. When females hunt alone, they are more likely both to catch prey and to keep their prize once the capture is accomplished. Meat is a much-prized resource to chimpanzees, and a prey animal is rarely consumed by only one individual, unless that individual was alone when the kill was made. For when a capture occurs, the other members of the group covetously and aggressively converge on the successful hunter. In this male-dominated society, a female, or a low-ranking male, is then likely to lose all or part of the carcass that she or he worked so hard to obtain. Even dominant males seldom manage to retain possession of all of their prey. They may be attacked for it and lose at least part of the kill to the attacker; or other chimpanzees will beg for meat, which is normally the only way in which lower-ranking individuals will manage to get any.

How the ultimate share-out is managed will often depend on the social relationships of the individuals involved, although how strong, tolerant, or satiated the possessor happens to be also plays a role. Thus, while the hunting itself may be a group activity among chimpanzees (though perhaps not quite in the sense in which we understand it), the sharing of the spoils is not. All this makes the entire hunting process quite distinct in chimpanzees and modern humans. Humans will typically plan a hunt; they will carry it out cooperatively; and the yield will be distributed among the members of the group according to certain rules, often after the carcass has been transported to a central place for division. Thus, viewed as a whole from initial intention through final consumption, chimpanzee and human hunting patterns are completely different. The common element seems to be not much more than knowing a good thing when you see one and knowing how to obtain it effectively.

If hunting is thus in its larger context qualitatively different among modern humans and chimpanzees, is there any more relevance to early human evolution in the fact that chimpanzees not only obtain such protein by this means, but that also—very occasionally—they scavenge it? After all, as we'll see later, scavenging is believed not only to have been an important early human activity, but to have retained much of its importance right up to a quite recent point in human evolution. Moreover, scavenging was once thought to have been—among primates—a uniquely human behavior; and seeing it practiced by another primate species forces us to view it in a new perspective. Or does it? Most of the rather few reported cases of chimpanzee scavenging have involved returning to the site of an earlier kill to finish off any leftovers, although scavenging of leopard kills has also been seen a few times. What's more, chimpanzees have been observed just as often to avoid animal carcasses they have happened upon. Dead animals are clearly not routinely viewed by chimpanzees as a desirable source of food, as they evidently were by our ancestors.

Of greatest morbid fascination among all chimpanzee killing behaviors are murder and cannibalism. The killing of adult members of one's own species is fairly unusual among mammals in general, and when it occurs, it is usually the by-product of other behaviors, such as male competition for females. Among chimpanzees, in contrast, murder (if we can use this term) has not only been observed, but appears to have been intentional and even premeditated. Vigorous aggression, exceeding that which typically occurs within local groups of well-acquainted individuals, is not uncommonly directed against strangers who are encountered. This may, of course, be conditioned by the fact that, while most primates travel in fairly compact and consistent groups whose other members will come to the aid of individuals who are attacked, the nature of chimpanzee social organization is such that a roving group of chimpanzees will often encounter a stray individual foraging separately from the other members of his or her community—and who is hence relatively defenseless. But certain kinds of attacks on adults are unique in their ferocity and intensity, and have usually been observed only in very specific—and apparently rather rare—circumstances.

The best-documented examples of such aggression come from Gombe, in Tanzania, where a large community of chimpanzees that ranged over a wide area split, presumably because it had grown too large for its territory or for proper social communication among its members. Jane Goodall's detailed account of the subsequent total annihilation of one of the two resulting groups by the other makes hair-raising reading, and one is inclined to believe her when she says of the (male) aggressors that "if they had firearms and had been taught to use them, I suspect that they would have used them to kill."

An observation that should make at least those of my gender

stop to think relates to the particular way in which chimpanzee groups are organized. Most primate groups are female-bonded; that is to say, the enduring "core" of the social group consists of females. Females born in a group normally stay within it, while maturing males leave to seek other company. Such transfer is how inbreeding is avoided in small populations. In contrast, among chimpanzees it is the females that transfer out, leaving the males to define and carry on the group identity. This is typical for humans, too.

Cannibalism represents if possible an even more grisly process than the one I've just recounted. It's such a touchy subject among humans that it has recently become fashionable to claim that all accounts of cannibalism in our species have been fabricated for political purposes. There can be little doubt, however, that cannibalism is or has been a behavior indulged in at least occasionally by human beings; and there is plenty of evidence that chimpanzees resemble us at least somewhat in this indulgence, although in all such cases so far observed, the victims have been infants. The killing of infant chimpanzees by adults was seen half a dozen times in a dozen years at Gombe and with equal frequency by Japanese observers at the site of Mahale, not very far away. It's been noted elsewhere as well. Infanticide is quite widely documented among primates and other mammals, and since it is often practiced by adult males after they have taken over the leadership of a group, it is often explained by sociobiologists as a device employed by males to eliminate the genes of others while promoting their own. Among chimpanzees, though, infants have been seen slain both by males and females, and in many cases the victim was eaten. What is most surprising for sociobiologists is that, at Mahale at least, in a large proportion of cases the infants were killed and eaten by males who could well have been their fathers. Almost as surprising, perhaps (especially since paternity is difficult to recognize), in a few cases the killers were female. And while in those cases where males killed infants the infanticide may be explicable as a by-product of attacks on the mother, in the instances of female infanticide the most obvious advantage gained by the killers was the flesh of the infant. It is possible, of course, that the motivation in such cases was vindictiveness toward the mother; but in one case, at least, the killer female embraced the victim's mother even as she was eating her infant. Such rare observations can be interpreted in a number of ways, some of which at least hint at the complex and murky motives that govern so much of human behavior.

One interesting aspect of chimpanzees' behavior that, like the killing and eating of members of one's own species, does not conventionally fall under the rubric of food seeking nonetheless dramatically illustrates the remarkable abilities of these apes to understand and exploit their external environment. This is self-medication. The forest is no Eden, and chimpanzees routinely suffer from diseases, parasite infestations, and so forth. There is by now plenty of evidence that certain plants are sought after by chimpanzees not for their nutritional or gustatory value, but for their curative properties, and at times when those properties are sorely needed. Thus the leaves of various plants, including species of the shrub *Aspilia*, are not always chewed but are instead sometimes swallowed whole, mostly during the rainy season when the prevalence of nematodes, intestinal parasites, is high. Besides containing known bioactive compounds including at least one antibiotic, these leaves are often coarsely textured and ream out anchored nematodes as they brush by the intestinal walls. They are eventually evacuated whole, carrying the parasites with them. Some leaves are manipulated in the mouth for quite long periods before swallowing, suggesting that the chimpanzees are attempting to absorb bioactive compounds through the membranes of the mouth. In Tanzania alone, more than a dozen plants consumed by chimpanzees have been identified as medicinal, and analysis has shown their ingredients to include antibacterials and agents of headache and stomach pain relief. Even though studies of self-medication by chimpanzees are in their infancy, it already seems clear that these apes are able to associate their feelings of physical distress with the healing properties of certain plant parts and with the fact that those items are most effective when ingested in a certain way. As far as we know, no primates other than apes and humans are able to make complex associations of this kind.

To the extent that it is reflected in flexibility of behavior, intelligence is a rather generalized capacity, and it's hardly necessary to demonstrate that chimpanzees' (or orangutans') food-seeking behaviors are similar to ours to suggest that these primates might not have an intelligence—or behaviors—that might be equivalent, even if not identical to, that of our remote ancestors. This suggestion has, indeed, been made. The chimpanzee specialist W. C. McGrew has compared the food-gathering tactics of chimpanzees with those of the Aboriginal Tasmanians, a now-extinct society of hunting-gathering humans who possessed a very simple material culture. Materially this culture was so modest, in fact, that it would have left little evidence in the archaeological record. McGrew has energetically argued that while the range of human foraging activities is far greater than those of chimpanzees, in certain areas there are significant similarities. Many of these, however, concern the use of tools, and that is a subject we'll pursue separately in a moment. Meanwhile, let's turn our attention to the social environment as a possible stimulus to increasing intelligence.

## *Intelligence and Society*

If you were to twist the arm of a doubtless unwilling anthropologist to tell you *why* humanity's high intelligence originated (or, as I would prefer to put it, why more intelligent species in our lineage succeeded), it's more likely than not that he or she would eventually mumble something about "our complex social milieu." Well, it's certainly true that, along with their larger and more elaborate brains, the higher primates exhibit a considerably more pronounced "social intelligence" than their closest relatives the lower primates. Even I, a longtime student and admirer of the most diverse group of lower primates, Madagascar's lemurs, find myself obliged to go along with this notion—despite my strong belief that lemurs' social complexity has been underappreciated, perhaps because their forms of communication are more foreign to human observers than are those of monkeys and apes.

There's no doubt, then, that as we move from lemurs, to monkeys, to apes, and finally to humans, the immediate social settings in which these creatures live, move, and have their beings become more complex: that, in effect, their social environment occupies an increasingly large proportion of their total environment. Among modern urban peoples, the social environment is often virtually the entire environment, with consequences we'll examine later. In view of all this, it's not hard to see why, while many primatologists have sought explanations of the overall organization of primate groups in the external environment—the minimization of predation, the most efficient exploitation of available resources, and so forth—those interested in the mechanisms that promote increased intelligence have inclined instead to look toward the complexity of interindividual interactions *within* social groups. And what they have found has led them to characterize the kind of intelligence required by such interactions as Machiavellian. The main reason for this rather unkind (or perhaps complimentary) designation is that life within social groups that—unlike bee colonies—are composed of individuals whose behavior is largely voluntary is hardly an idyll of harmonious cooperation. Rather, individuals are in competition with each other for a wide range of resources that include such essentials as food, status, and sex; and it's here that the Machiavellian aspect of intelligence comes in.

The reason for this is that in relatively stable and hierarchically organized groups such as those of most higher primates, social success often depends less on your raw capacities—your strength, charm, aggressiveness, or whatever—than on who you know. Most Old World monkeys and apes, for example, live in societies with distinct if fluid dominance structures that are often linear (A displaces B displaces C, and so on). Early interactions, and usually the status of the mother, determine where an individual initially falls in the hierarchy. And this start in life is important, for high status often brings with it preferential access to resources, while low status often correlates with such disadvantages as high parasite loads and spatial peripheralization that brings with it greater vulnerability to predation. Labyrinthine though our own industrial society may be, you'll immediately appreciate where we humans fit into this picture.

But status is subject to change; and upward mobility (from whatever position is determined by that of the mother and her consorts) depends on many factors—among which strength and aggressiveness, for example, are not necessarily critical. More important, it turns out, are the alliances formed by each individual throughout life. And alliance building implies politics, however rudimentary, which is what brings us back to Machiavelli.

The literature on higher primates in their natural habitats is by now replete with observations on how, for example, lower-ranking males of this or that species managed through coalition forming to gain access to sexually receptive females that alpha males were attempting to monopolize, or on how two lower-ranking animals acting together were able to displace a higher-ranking individual from a favored feeding resource—and on and on. Interindividual relationships are complex among members of all higher primate societies, and linear status relationships in those societies are constantly blurred by shifting patterns of alliance among individuals. The aspect of such social manipulations that has particularly engaged the interest of those concerned with assessing intelligence is deception—a tactic that is apparently widely employed among higher primates. Field observations of deceptive behaviors are necessarily anecdotal, but they are telling. For example, an infant baboon may pretend to its mother that it is being harassed by a larger juvenile so that it can escape with the latter's food as the mother gives chase; or a female may use a visual obstacle to obscure from a dominant male that she is being groomed by a subordinate male. Such duping of individuals into false beliefs suggests, significantly, that the deceiver is able to attribute states of mind to the dupe.

This notion has been most fully explored by cognitive psychologists working with captive animals under more controlled conditions than are normally possible in the field. Let's take one example.

In a famous study aimed initially at investigating spatial reasoning and memory in young chimpanzees, one individual, named Belle, was shown where a pile of food had been hidden in a large enclosure. She was then returned to her peers, and the whole group was let into the enclosure together. The first few times this was tried, Belle mostly led the group to the food, which was duly shared by all. But then a stronger young male, Rock, began to monopolize the food that Belle uncovered. In response, Belle refused to reveal the food's whereabouts when Rock was nearby, sitting on it until Rock had left the scene. But Rock began to observe where Belle was sitting and started to push her aside, recovering the food. Belle countered by sitting away from where she knew the food was and not moving toward it until Rock was looking in the other direction. Rock's response was to look away until Belle started to move; sometimes he would even wander away. But then he would turn around suddenly and look back as Belle reached the food—which he then appropriated. In some trials, Belle took to leading the group in the wrong direction and as Rock foraged for nonexistent tidbits, running back to the food cache. Occasionally an extra piece of food was hidden away from the main cache. Belle took to leading the group to this, then rushing over to the main food pile while Rock was engaged in eating the single piece. Eventually Rock caught on to this, too, and began to ignore single food pieces. At this point, the frustrated Belle could only respond with temper tantrums.

Psychologists have been assiduous in trying to distinguish cases of "intentional deception" from those of "unintentional deception," where apparently deceitful behavior results simply from the employment of devices that experience has shown to be effective. For while the learning abilities shown by "conditioned" behavior of this latter kind may be impressive, they are not evidence of one individual's ability to read the mental states of another. Many of the behaviors shown by Belle and Rock in the escalating sequence just described might be explained by conditioned learning, as could the baboon behaviors I mentioned earlier; but once Rock had begun to pretend a lack of interest in order to defeat Belle's own deception, it is pretty clear that he had understood the logic of Belle's actions and had anticipated what she would do next. Rock had, in other words, formulated an image in his own mind of what was going on in Belle's. This ability has been confirmed by a number of other observations in apes, some of them under natural conditions, but has yet to be identified unequivocally in monkeys. Thus, while manipulative social behaviors of various kinds are routine among higher primates, it seems that humans share only with the great apes the ability to enter the minds of other individuals and to predict their behavior from the knowledge thereby gained.

If the ability to assess the thoughts of others is present in some form among apes, how about the knowledge of *self?* After all, it has long been argued that our only true knowledge is of ourselves: that our knowledge of the outside world is fabricated from our own constructs and filtered through our perceptions. The matter of self-knowledge is an extremely difficult one to approach, even in humans, and so far the most successful attempts to address it in other primates have employed the rather rudimentary tactic of using mirrors. When presented with mirrors, monkeys have proven adept at using their reflective qualities; they have, for example, used them to see around corners and identify other monkeys—an ability that capitalizes on the appreciation of visual geometry shown by examples of visual deception. But they rarely if ever seem to get the idea that when facing mirrors straight on, it is themselves that they see reflected. Great apes, on the other hand, do manage to figure out what's going on (after an expected initial period of confusion, which is also seen in adult humans unfamiliar with mirrors). Thus, for instance, when spots were painted on chimpanzees' faces under anesthetic, most of the subjects later wandered around unconcerned until they caught sight of themselves in a mirror. They then tried with its aid to pick off the paint. And once they have recognized themselves in mirrors, chimpanzees regularly use the reflections to explore their faces and bodies, looking this way and that, and pulling at various features to see what results.

It has been argued convincingly that the evident ability of apes to recognize their own reflections implies that the individuals concerned possess a concept of self; for if they had no such concept, they would lack any means of correctly interpreting the image. And it has also been argued that knowledge of oneself and one's own mental states is an essential prerequisite of understanding the thoughts and feelings of others. Such interpretation is, of course, a very different process than the simple anticipation of what response one's own actions might elicit in others—even though the behaviors resulting from the two processes might be difficult or impossible to distinguish in practice. And it must be admitted that apes exploit their images in mirrors far less comprehensively than humans do. It has been noted, for example, that apart from the removal of unfamiliar markings, apes make no attempt to modify their images, even in ways that might make them socially more successful. There is, for example, no hint of a desire to "improve" the reflection and its subject, as humans in all societies (with or without mirrors) do—by cutting or coiffing the hair, for instance, or by embellishing themselves with jewelry or makeup.

The list of behaviors in which apes—and apes alone—at least partially resemble humans could go on and on. But it's self-evident that no observation on nonhuman primates will ever document a *sufficient* cause of human intelligence—however necessary some of the capacities revealed may have been as way points in its acquisition. Humans are still humans, after all, and chimpanzees are still chimpanzees. Higher primates started out with equal potential for all, in the form of a single common ancestor (which could not, itself, have been greatly different from its closest relatives). The various bifurcating higher primate branches that descended from the common ancestor simply capitalized—differently, and over a long period—on the particular quality that ancestor possessed predisposing it for increased intelligence. As we'll see in more detail later, natural selection is itself subject to random events—it can only work on situations presented to it—and it most certainly is not an optimization mechanism in an engineering sense. Modern humans are the result of a unique series of evolutionary occurrences that depended not only on the attainment of particular heritable capacities, but on those other capacities that had already been acquired by their lineage when selection interposed itself.

This aside, it's certainly true that, in the larger scheme of things, the behavior of apes anticipates that of humans in a variety of respects. Indeed, as is readily apparent even from the very sketchy comparisons above, the perceived cognitive gulf between us is tending, within limits, to narrow. To revert to an earlier theme, though, it is pointless to seek the *causes* of our unique intelligence in similarities between our capacities and those of apes and monkeys. What ape-human comparisons can do is both to help establish the background out of which our remarkable mental abilities emerged and to allow us to gauge the extent of the similarities and differences between us. This is essential if we wish to evaluate our own uniqueness—and in the process, perhaps, it will give us pause when we consider how our species tends to treat its nearest but evidently not dearest. With these things in mind, let's now turn to the two human behaviors, toolmaking and language, that have attracted most attention from anthropologists, cognitive psychologists, and others concerned with how we got to be the way we are.

## *Toolmaking and Tool Using*

According to James Boswell, it was Benjamin Franklin who coined the term "Man the Toolmaker" to denote the central capacity that sets us apart from the rest of nature. And although more recent knowledge has shown that we are not alone in using tools, or even in making them, it is nonetheless true that humans today, worldwide and irrespective of means of subsistence, stand out in being totally dependent on tools to sustain their ways of life. Franklin really was on to something. As we'll see in a later chapter, though, tools—at least of the kind that preserve in the archaeological record—are a relatively recent innovation in our lineage. For at least the first half of human existence—broadly defined—we have no evidence whatever that durable tools were made. We may, however, quite confidently infer that perishable materials such as sticks were used by our ancestors as implements well before hard stone tools show up in the archaeological record at about 2.5 million years ago. One reason why we can be pretty certain about this lies precisely in our recently acquired knowledge that we are not alone as tool users in the living world. Tool use, it turns out, is not even restricted to primates: birds, otters, and even dolphins have been seen to use objects as intermediaries in carrying out various activities; and probably close enough examination of many mammals would allow us to come up with an example or two of what would in the widest sense have to be called tool use. Yet Australian finches using thorns to extract tasty grubs from deep inside trees, or even otters using stones to crack open crab shells, seem to be indulging in isolated behaviors that represent learned responses to very specific problems. For "intelligent" approaches to tool using, we have to turn to primate examples.

Surprisingly, once we narrow our definition of tool use, we are left with rather few candidates, even among our primate relatives. As you'd expect, the great apes are the leading contenders, but when we move farther afield, the pickings are slim. Indeed, the South American capuchin monkey—the organ-grinders' favorite—is about the only other spontaneous primate tool user. Capuchins have been observed in the wild using stones to crack nuts, and one has been seen belaboring a snake with a stick. Most observations of capuchin tool use have, however, emerged from laboratory settings, and recently a particularly interesting captive study helped place these observations in perspective. In this study, capuchins were given the opportunity to use a stick of appropriate diameter and length to poke peanuts out of a transparent tube. Eventually most of the monkeys who took the test got the idea (and the nuts); these individuals were then given a variety of sticks, most of them inappropriate, and invited to try the task over. Most of the monkeys were again successful, although only after much trial and error during which immense amounts of time were spent poking around with sticks that experience should rapidly have shown were the wrong length, or width, or whatever. This hardly demonstrated an understanding of the properties that made a particular stick into a functional tool.

Undaunted, the experimenters then made the problem a little more difficult. A depression was made in the underside of the tube into which a peanut would fall—and be lost to the monkey—if it was pushed in the wrong direction. Only one of several subjects succeeded here, learning consistently to push the peanut away from the pit and thus out of the other end of the tube. But even this individual had evidently failed to understand the significance of the pit itself. For if the tube was rotated so that the pit would no longer trap the peanut, this champion capuchin would still push the food in the previously successful direction, even when this meant pushing it much farther. The monkey had not understood the significance of the peanut trap itself. Putting all of these observations together, it seems that the capuchins were pretty good at trial-and-error learning but had hardly mastered the principles of the problem. In other words, they showed little comprehension of cause and effect in their tool use.

Not so chimpanzees, for which we now have a lot of information on tool use not just in the laboratory but in the wild. Chimpanzees use a lot of different objects for a lot of different purposes, only a few of which there's space to mention here. Perhaps the most impressive of these tool-using behaviors is shown by West African chimpanzees, which crack nuts open by hitting them with a "hammer" stone (logs are sometimes used, too) while they rest on a large "anvil" stone—though by doing so they don't intentionally modify anything except the nut. The nuts won't break open if they are bashed while lying on the soft earth floor of the forest, so anvil stones are important; some show evidence of repeated use over long periods of time. Hammers are often in short supply, too, and the chimpanzees will if necessary carry them over half a mile to a source of nuts, though these distances are often the shortest possible, reminding us of the "mental maps" discussed earlier. Nut cracking (like other behaviors analogous to it) does not occur at all sites at which the necessary raw materials for this activity are found, and this reinforces other evidence that many if not all chimpanzee behaviors of this particular category are "culturally" conditioned: that they are voluntary behaviors transmitted by learning from generation to generation only in certain populations. Necessarily, they are dependent on the general capacities of the species concerned; but they are not species-wide behavior patterns.

The most familiar examples of tool use among chimpanzees come from Tanzania (where nut bashing is not practiced, although the nuts and rocks are there). These activities are, of course, the widely publicized termite fishing and ant dipping. Here thin twigs are chosen, stripped if necessary, and then inserted into termite mounds or ant nests. The intruding probe is attacked by guard insects, which continue to cling to it when the twig is withdrawn and thus provide a tasty morsel for the dipper. Twigs of different kinds are selected for different purposes, and recent observations reveal that stouter branches are used as levers or to dig out honey from bees' nests. Significantly, twigs are not necessarily discarded when they become bent or frayed; as long as they can, chimpanzees will usually break off the end of such a tool to "refresh" it and will continue using it as long as such modification is possible. Chimpanzees have also been observed to break off branches for use in hooking in fruit from otherwise inaccessible tree limbs, for attacking potential predators, and for expelling the occupants of holes in trees. Branches are also brandished to enhance the effectiveness of aggressive displays, and rocks and sticks are thrown in attempts to intimidate competitors or predators.

Many other items, mostly foliage of one kind or another, have also been seen to be employed by chimpanzees for various purposes. Banana leaves are used as "umbrellas" during heavy rain; leaves are used by chimpanzees to clean themselves after defecation or to wipe blood from a wound; chewed masses of leaves have been used to sponge up water which was then squeezed into the chimpanzee's mouth. With every year the list of external objects spontaneously put to use by chimpanzees in the solution of specific problems grows longer; my wish here is not to make an exhaustive inventory of these, but to make the point that chimpanzees are capable of selecting a wide variety of objects, and in some cases of modifying them, in the pursuit of diverse goals. This does not mean that chimpanzees are toolmakers (or even tool users) in the sense that modern humans are—clearly, they are not—but it shows that chimpanzees are capable of forming a mental picture of what attributes some simple tools, at least, need to have to accomplish a particular aim.

Such chimpanzee proclivities have attracted the attention of many who study these primates in captivity. Among the earliest such investigations was one carried out by Wolfgang Köhler in the 1920s. One of his chimpanzees, Sultan, eventually figured out how to join two sticks together to make a single rod long enough to rake food to within reach. Although it was later pointed out that chimpanzees—with nothing better to do—will sometimes join sticks in the absence of a problem, there's no doubt that Sultan understood the relevance of this activity to the situation at hand. Köhler's laboratory investigations have been followed up by many others that confirm that Sultan's abilities are not unusual among chimpanzees; and the most interesting of these subsequent inquiries has focused on a question particularly dear to those concerned with the origin of our own human capacities: the abilities of chimpanzees to make *stone* tools. Or rather, on such capabilities as shown by *a* chimpanzee, Kanzi, of whose diverse achievements we will hear a great deal.

So far when I have quoted studies on chimpanzees, I have been referring to *Pan troglodytes*, the inappropriately named "common" chimpanzee. Kanzi is a bonobo, belonging to the species *Pan paniscus*, often referred to equally inappropriately as "pygmy chimpanzees." Bonobos and chimpanzees are closely related but show intriguingly different behavior patterns that include a lack of male dominance among bonobos. Kanzi started his academic career as a star in communication experiments, about which we'll learn more in a moment; but for the present we'll look at his aptitudes as a stone tool*maker*—remembering, of course, that this is an activity in which neither chimpanzees nor bonobos indulge in the wild, for even the nut-cracking West African chimpanzees do not modify the stones that they use for this purpose. Nonetheless, for reasons that should already be abundantly clear, chimpanzees and bonobos have attracted a lot of attention as possible "models" for the behavioral capacities of very early humans.

The earliest human stone tools consisted of sharp flakes chipped off small cobbles. Not very impressive to look at, perhaps; but these very rudimentary implements, especially when made from the most appropriate materials, are very efficient cutting devices that served humans well for millions of years. Producing such useful objects is, however, tougher than it seems—as anyone who has tried to make them will tell you, doubtless rubbing their sore and scarred hands as they do so. The trick lies in striking the cobble with a hammer stone at precisely the right angle, and thus in understanding—at some level—the properties of the material that will cause it to break in the desired way when struck. Knowing that chimpanzees spontaneously modify objects to achieve various goals, the experimental archaeologists Nick Toth and Kathy Schick decided to see whether a chimpanzee might be capable of learning the principles involved in stone tool production. An orangutan at England's Bristol Zoo had already been taught to strike a flake from a preshaped stone core that could then be used to cut a cord holding a food reward out of reach; and although bonobos (like orangutans) have not been observed to make tools of any kind in the wild, Kanzi, with his demonstrated abilities in other areas, seemed to be an ideal candidate for a test of this kind. The procedure involved getting Kanzi to want to produce a cutting flake to gain such a food reward; showing him how such a flake could be obtained; and providing him with a variety of rocks from which tools might be produced. No coercion was involved; Kanzi always had something to do other than indulge in the desired activities.

Kanzi soon got the idea of using an existing flake, produced in front of him by the experimenters, to obtain a food reward. Next he rapidly learned to select a flake capable of cutting the cord from an assortment of rocks provided to him. At this point, he was simply left with the inaccessible reward and with the raw materials needed to obtain it. Although he did not pursue this goal very energetically, Kanzi eventually produced several flakes from a rock core and cut the cord with the largest of them—the first he had produced. In the process, he spontaneously bashed the core against an "anvil" stone on the ground—something he hadn't been shown how to do. Over a period of months Kanzi was able to replicate this success on a number of occasions, although he was not using enough striking force or enough directional control to be a reliable tool producer. Months later yet,

Kanzi realized that he could produce flakes by throwing a core energetically at a hard floor. This became his preferred method, and he only once produced a flake by hitting a core held in his hand. He was then moved outside, where the ground was soft, and found himself unable to use the throwing technique. After this he refined his two-handed technique, bashing harder and more accurately with the hammer stone yet still not producing flakes with any efficiency. His final advance was to discover—again spontaneously—that he could produce fragments quite reliably by throwing a cobble against another lying on the floor.

What did all this add up to after a year or so of experimentation? Clearly, at one level Kanzi learned well by imitation—something already strongly evident in other great apes (orangutans in particular, it seems), who often spontaneously—though, like Kanzi, inefficiently—mimic human behaviors, even quite complex ones, such as washing clothes or lighting fires. But Kanzi never really figured out what was going on; he was evidently imitating the actions of his human teachers, without worrying about the exact nature of the results he obtained. Of course, Kanzi did get the basic notion of using percussion to fracture rock—and even worked out new ways of doing it; but he never managed to reproduce reliably the result toward which the experimenters were coaching him. Whatever the technique he used, the results were pretty much the same: he produced rather small flakes, mainly of the kind that might be produced by random battering of stones (on a streambed, for example).

Kanzi thus failed the ultimate test: he showed no insight into the properties of the materials he was working with or into the principles by which such materials could be fractured to produce results comparable to those the experimenters had shown him. Of course, Kanzi's rather dismal performance might have resulted from lack of motivation rather than from lack of innate ability; but given the length of the experiment and the fact that for a while his performance did improve, this is probably not the whole story—though, as I've hinted above, two-handed stone toolmaking is painful for beginners, and some aversive conditioning to using adequate force might be expected; you can hardly blame Kanzi for preferring to procure flakes by throwing. What's more, it's not possible to know for sure whence Kanzi's stone toolmaking deficit came: whether it lay in his central cognitive abilities, in his neuromuscular coordination and control, or in both. And we don't know whether, as far as these particular capacities are concerned, Kanzi is a typical bonobo. Perhaps he's just a klutz in this respect, however remarkable his other abilities.

As work with Kanzi and other apes proceeds, it's possible that some of these questions will be answered. For the time being, though, it's important to note that the earliest hominids who made identifiable stone tools clearly possessed both the central and the peripheral mechanisms necessary for this activity. But perhaps most importantly, they *invented* efficient toolmaking from materials they consciously chose, and this activity became an important component of their behavioral repertoire. Cognitively, it's clear that they were significantly different from any living ape. It's worth emphasizing here, however, that when interpreting the experiment with Kanzi, we have to be extremely careful not to look upon him or any other ape as some kind of inferior human, even though they may in some very limited respects reflect a stage of cognitive development through which very early prehumans passed. Like all species, ours does business differently even from its closest relatives, and intelligence expresses itself in a variety of ways. Kanzi, we must never forget, stands at the end of a long lineage that has been diverging from humans for exactly as long as humans have been diverging from it.

## *Apes and Language*

Perhaps no human ability has attracted more attention than our capacity for language. In turn, the mysterious nature of this capacity has long excited speculation about language abilities in other species. Two centuries before Darwin, for instance, Samuel Pepys was sufficiently impressed by the behavior of a baboon brought back to London by a sea captain that he wrote in his diary: "I do believe it already understands English, and I am of the mind it might be taught to speak or make signs." A hundred years later the Scottish jurist and philosopher James Burnett, Lord Monboddo, evidently felt something similar on reading travelers' reports that orangutans were half human and principally distinguished from ourselves by lacking language. Early rumors have, of course, proven unfounded that in the wild, orangutans are bipedal, live in shelters, bury their dead, and so forth; but Monboddo was clearly correct in his belief that language is "necessarily connected with an[y] enquiry into the original nature of Man." The learned and intellectually refined Monboddo was a fountain of unusual ideas and was written off as a crank by many of his contemporaries; but few in later years have taken issue with his notion that language is central to our unusual abilities. Language is intimately tied up with our complex symboling capacities, and is, indeed, the medium through which we explain those capacities to ourselves. Universal among modern humans, language is the most evident of all our uniquenesses: the one in the absence of which it is least possible for us to conceive of humanness as we experience it.

Interindividual communication, on the other hand, is a different matter, since to one degree or another it is characteristic of all complex life-forms. Among primates, communication between individuals is particularly elaborate, although *vocal* communication clearly predominates only among higher primates. In turn, among nonhuman higher primates, vocal communication is particularly well developed in the apes. This suggests that there do exist functional levels of complexity of vocal communication among living things, with humans at the highest level and the apes on the rung below. Two questions then arise. First, do these levels of complexity form a continuum, in which case might it be possible that the vocal communication of apes represents a prelinguistic stage through which our human ancestors passed? Well, it does seem reasonable to suppose that at some remote point in our evolution, our ancestors communicated in ways that broadly resembled those of apes living in complex groups. But whether such behavior can be regarded as prelinguistic (except in the strict sense of time) is another matter entirely. Hence the second, and more subtle, question: Do apes possess any of those aspects of cognition that might have predisposed our ancestors to acquire language? We'll tackle this issue in a moment; but before we do so, we need to look at how apes communicate in their natural environments.

Once again, it makes most sense to use chimpanzees as our example, if only because they live in the most extensive groups known among apes, have been closely studied for many years in the wild, and have been the subjects of most laboratory investigation. Like humans, chimpanzees have a rich repertoire of postures, gestures, and facial expressions that they use to convey intentions, desires, warnings, submission, and so forth. Indeed, different nonverbal "dialects," involving a wide variety of signals, have been identified in various chimpanzee populations. And it's only fair to note that we may underestimate the importance of such elements in our own communication because they are overwhelmed by language—a much more refined and precise method of conveying (and, more significant in this context, of explaining) the same things. Nonetheless, if we wish to understand the substrate out of which language may have emerged, it is clearly at vocal communication that we need to look.

Studies of wild chimpanzees have revealed that they make a wide range of calls: well over thirty different vocalizations have been identified and named. Even this large number presumably underestimates the true complexity of the system because all thirty-plus sounds actually seem to form part of a graded series, and subtly different calls may be lumped under a single name. What's more, combinations of different calls are regularly used by chimpanzees, and this adds a further dimension of complexity to the system. Nonetheless, in the field situation—where the meaning of any sound has to be inferred from indirect evidence—observers have been reluctant to impute abstract significance to any of the calls that have been identified. Certain calls are characteristic of particular activities, but none of them appears, for example, to be communicating instructions. Even when indulging in such complex behaviors as cooperative hunting, individual chimpanzees seem to take their cues from the movements of others, rather than from any sounds they might emit.

What chimpanzee vocalizations do appear to reflect very closely, on the other hand, is the emotional state of the vocalizer—information that is critically important in a society based on complex interindividual relationships. Indeed, Jane Goodall believes that vocalizations are so closely tied to emotional states that "the production of a sound in the *absence* of the appropriate emotional state seems to be an almost impossible task for a chimpanzee." Even among chimpanzees, sound production appears to be controlled in the brain by the ancient structures of the limbic system and the brain stem, which we'll read about shortly and which are involved in emotional response. The "higher" centers of the brain do not appear to be much involved. This is a far cry (sorry!) from language as we humans know it, which is initiated in those higher centers (the cerebral cortex) and is dependent on production and interpretation of sounds in isolation from the emotional states of the speaker and hearer. It is also dependent upon rules of grammar, syntax, and so forth that are totally absent from the sound combinations chimpanzees make. So, no. Not only do chimpanzees not have language; they don't even have an incipient form of it.

To make this point even clearer, let's look at human language for a moment. As you might expect, this complex construct has so far resisted definition to everyone's satisfaction. But more or less everyone agrees that it contains a number of interrelated elements, which are probably also interdependent. First, language assigns wholly arbitrary meanings to particular sounds. All English speakers know what a house is, but the word *house* has no association with its subject other than that we learned it was so. There's no way a monolingual Spaniard could make that association. Second, words are not limited to denoting concrete objects; a capacity for *symbolic reference* also permits them to refer to qualities or entities that exist only in an abstract sense and not necessarily in the present. Third, the order in which these arbitrary symbols are combined conveys its own meaning, independently of those symbols themselves: *man paints house* describes a sight that is familiar every summer, while *house paints man* is just puzzling. Consider the number of words in even a limited vocabulary; the number of words in an average sentence, let alone a long one; and the number of ways in which those words can be arranged, and you'll quickly see that language is virtually infinite. Of course, not all word arrangements will make sense, as in the example I've just given. Not all need to; but you do need a code to identify those that *are* meaningful and to understand just what they are designed to convey. This code is what in English we call generative grammar, and it is just as arbitrary as the meanings of the words whose arrangement in a sentence it governs. To use language at all, then, you need to possess both a vocabulary and a command of the grammatical code to help you string those words together such that your hearers, possessing the same arbitrary learned knowledge, will correctly interpret them.

All human societies possess language, and there are said to be as many as six thousand languages actively spoken in the world today, though most are hanging on by a thread. The vocabularies from which these languages are built can vary wildly, as can the vocal gymnastics demanded and the logic that governs their use. Nonetheless, all human individuals possess the ability to learn any language—if they start early enough. The way in which children acquire all languages—for example, the order in which they master their complexities, the kinds of error they make, and the overall speed of learning—strongly suggests that in our species there is an innate and generalized ability for such acquisition. This ability correlates with the maturation of the brain itself, diminishing strongly after the first decade of life. None of this is to say that the environment in which language is learned does not play an important role; evidence increasingly shows that it does and that stereotyped behavior on the part of parents and others helps the process along. In all of this, the organization of our brains clearly plays a pivotal role, but it is far from the only thing. The ability to produce the sounds associated with articulate language depends on the form of the vocal tract, so the development of the vocal apparatus is also critical to successful speaking. We will get back to this point; meanwhile, it's sufficient to stress that language is not simply something we invented on a whim, but is intimately tied up with our physical evolution. Now let's briefly return to the apes.

The last few pages will, I hope, have convinced you that chimpanzee vocalizations in the wild serve a qualitatively different purpose from the sounds associated with human speech. But could there, nonetheless, be something in the chimpanzee cognitive apparatus that might predispose these apes to language acquisition? Some cognitive psychologists have thought so, to the extent of investing considerable time and effort in attempts to demonstrate it. The first of these attempts dates back to before World War II and involved intensive efforts to train a young chimpanzee to form words and use them. After years of this, the chimpanzee had only learned to "speak" three words that a receptive audience might recognize as such and to recognize a few more spoken by her trainers. What's more, she also showed a bewildered incomprehension of the significance of word order. Later researchers, yielding to the obvious deficiencies of the chimpanzee vocal apparatus (for carrying out a human activity), have taken various different tacks. Some of these involved attempts to teach chimpanzees American Sign Language (designed for deaf humans, with all the complexities of spoken language). In one such case, it was claimed that a chimpanzee had mastered more than 130 signs over a four-year period and had mustered them into simple phrases. Using another approach, researchers mimicked language with colored plastic chips intended to stand for particular words and persuaded one chimpanzee (out of four) to identify them and arrange them in sequence to obtain rewards. More technophilic investigators provided a chimpanzee with a sort of keyboard, connected to a computer, on which a variety of abstract and color-coded symbols represented desirable objects and concepts, such as "please." The chimpanzee had then to press the keys in learned sequences in order to obtain the indicated incentives.

There's no need to go into a lengthy description of these experiments and various others that were conducted both with chimpanzees and gorillas, enormous effort being poured into training the subjects to respond "appropriately." They have been covered exhaustively in many books and articles, and it's an indication of the intrinsic fascination of the exercise that a blast of publicity greeted each ape's announced success, with headlines declaring that one or another had achieved the ability to communicate with humans. The truth, alas, is considerably more prosaic. One of the most carefully designed efforts to teach an ape American Sign Language (in a suitably modified version) resulted in the conversion of the investigators from enthusiasts to skeptics, and things went downhill from there. It turned out that in their anxiety to show that apes could acquire skills related to language, experimenters had vastly overinterpreted their subjects' responses. Close examination of the experiments revealed excessively optimistic readings of many of the observations. One reappraisal, for example, reduced the 130 signs mentioned earlier to a mere 25. The subjects certainly learned to recognize certain basic signs made by their trainers; but they tended to return them with similar gestures taken from their natural repertoire. They learned to recognize some elementary words spoken by humans; but then, so do sheepdogs. They even managed to identify and respond to word combinations (the command "take the vacuum cleaner outside" might result in the appropriate behavior); but this shows no more than that they were able to make associations between two words (they could ignore the rest of the sentence and still perform perfectly well; they could hardly bring the outside to the vacuum cleaner). But in no case could it be shown even to the investigators' complete satisfaction that an ape had acquired any understanding whatsoever of grammar or syntax, even after the most extensive training. Moreover, there was no evidence of a learning curve. Whereas a human child rapidly acquires the ability to piece together longer and longer sentences, the apes tailed off rapidly after getting the idea of putting a few signs together. In sum, there was no suggestion at all that the apes ever really understood what was going on. Apes learn to make requests; but unlike even very young children, none has ever even tried to initiate a conversation.

This is, of course, putting the results of these experiments in the most unfavorable light possible. On the positive side, the tests demonstrated quite impressive learning abilities in their subjects. They showed that the apes could make referential associations: that they could grasp the notion that verbal, gestural, and other symbols can be made to stand for concepts or objects. And they showed that their subjects were able to associate two concepts embedded in a longer verbal command. Hope thus springs eternal, and even as I write, our old friend the bonobo Kanzi is still engaged in an "ape language" experiment. Most such experiments have been carried out using behavior modification by reinforcement (reward); but Kanzi's trainers, initially concerned with attempts to train Kanzi's mother, noticed that the infant—who just happened to be around—picked up some symbols purely by observing her hit them on a keyboard. Evidently he could combine learning by imitation with at least some referential capacity. Abandoning structured situations, his caretakers thus began simply to point to the keyboard symbols as they discussed items in Kanzi's environment, and he rapidly learned to recognize the meanings of quite a large number of such "words." It's even claimed now that Kanzi can organize the symbols into a "sentence" containing three symbols and can recognize the significance of the order in which such symbols are presented. It's hard, though, to demonstrate that these "sentences" are anything more than successful formulas, rather than properly parsed expressions. And Kanzi has made considerably slower progress than any human child would have done. What gains Kanzi will continue to make as the experiment continues are anyone's guess, although mine is "not many." And, as apes do, he may also become more fractious as he matures, bringing the experiment to an end. Kanzi has been touted in magazine articles and a popular book as the ape "at the brink of the human mind"; but at best he has so far only demonstrated an ability to do what other chimpanzees have also been (laboriously) trained to do, if a tiny bit better. Kanzi is no more articulate, even with his keyboard and dedicated trainers, than he is proficient as a toolmaker.

Chimpanzees, then, do not have language. Nor do they, even through the heroic and persistent efforts of human trainers, acquire even a rudimentary form of it. Further, it is hard to demonstrate that they possess any cognitive abilities that could be called prelinguistic. Human beings are truly unique in having language and in possessing the apparatus that permits them to acquire and express it. Lest I be thought a chimpophobe, let me hasten to add that there is no reason why apes *should* have an ability to acquire language. The problem lies with us. Human beings regard themselves as having reached an evolutionary pinnacle—no, *the* pinnacle—and they like to emphasize this lofty position by viewing their close relatives as lying farther down the very slope they have so successfully mounted. But viewing evolutionary success as analogous to climbing a ladder distorts our perspective. There are many ways of doing business in this world, and apes have their own ways just as we do. Vocal communication in apes serves a different set of purposes from vocal communication in our species; and it is totally wrong to see the apes' version as no more than an inferior form of our own. If we think in terms of pinnacles at all, we have to recognize that there are other pinnacles in the world than the one we occupy: one, indeed, for every one of the world's millions of species.

This having been said, it is presumably true that modern human articulate language—which involves not only central neural mechanisms but an external speech-producing apparatus as well—cannot have sprung fully fledged from the larynx of a totally inarticulate species. Fortunately, few would raise here the famous paleontological problem (actually it's a nonproblem, but it would take a long digression to explain why, and it's been done many times already) of intermediates: how can a halfway flier, or biped, have operated as a smoothly functioning being? There are evident advantages to increasing sophistication of communication at every stage along the way from inarticulateness to language, all of them involving coordination between central and peripheral mechanisms. Few who accept that we evolved see any difficulty with the notion that our immediate predecessors possessed vastly better powers of vocal communication than our remotest ancestors did, even if those powers did not equal our own. The problem lies in envisioning exactly what those intermediates were; for the fact is that humans find it difficult if not impossible to imagine things that lie beyond their own experience or that cannot be extrapolated from it.

Thus you *know* your dog is "thinking" something when he looks at you with those big eyes; but what on Earth is actually going on in his mind—if he has one? Quite simply, you have no idea whatsoever what is happening inside his head, and you have to satisfy yourself with anthropomorphic constructs. Perhaps he is adoring you, or reproaching you, or wants to go outside, or maybe he is just bored and has nothing better to do. Whatever you tell yourself, though, your explanation will have much more to do with your own experience than with the dog's. Adept as you may be at reading the minds of members of your own species, you simply cannot imagine the dog's actual state of consciousness. Similarly with language. What are the possible intermediate conditions that lie between, say, chimpanzee vocalizations—the complexities of which we will never fully comprehend and which, anyway, have doubtless diverged from those of our common ancestor—and fully formed articulate human speech? What *could* they be? We can speculate; but, quite frankly, I doubt we'll ever know for sure. What we *can* reasonably surmise is that there are advantages even to small increments in the efficiency of communication among individuals in complex societies—especially societies that, like those of all humans, depend so strongly on communication and coordination among individuals in their economic lives. Language itself, however, with all the mental apparatuses of abstraction and association that it involves, does appear to represent a quantum leap away from any other system of communication we can observe in the living world. For it is intimately tied up with symbolic (as opposed to intuitive) reasoning, a matter to which we will return later.

Apes, then, have quite complex behaviors, societies, and modes of communication. Individual apes have distinct personalities and emotional needs, and they are capable of extreme suffering through social deprivation. To what extent they provide an adequate model for early hominid abilities and communication at any stage—except possibly the very earliest—is, however, more debatable. The cognitive gap between them and us has been narrowed somewhat by studies of the kind I've mentioned, but it is far from closed—and obviously never will be. Apes apparently cannot plan, have no capacity for abstraction, and only in the most rudimentary way do they associate past experiences as a guide to future actions. They do not display "generativity," the capacity that allows us to assemble words into statements, or ideas into products. This is clearly reflected in their lack of linguistic abilities, while our own language skills intimately reflect this capacity.

At best, then, the living apes provide a distant glimpse of the starting point from which our ancestors set out. They thereby define, if only approximately, the far side of the behavioral gulf that was bridged over several million years by our fossil ancestors. How and when this bridging was accomplished is the subject of chapters 4 and 5; but meanwhile, let's look briefly at the remarkable structure upon which our unique capacities depend.

## *The Human Brain*

Behavior is ultimately the product of the brain, the most mysterious organ of them all. Exactly what it is about our brains that leads to our extraordinary consciousness remains obscure; but we can certainly learn much from sketching the contrast between our brains and those of our closest relatives.

The first thing that springs to mind when we think about the modern human brain is its large size. Of course, we don't have the largest brain around; but then, we'd expect elephants to have much larger brains than us, for exactly the same reasons that they have bigger hearts, livers, and lungs. What's more surprising, perhaps, is that we don't even have the largest brains in *relative* terms: the South American squirrel monkey, for example, has a brain that weighs in at 3 percent of its body weight, while the human brain accounts for only around 2 percent of human body weight. Let's not be misled by this, however; for it simply reflects the fact that, among the huge spectrum of mammal species, brain size increases more slowly than body size in general—and most squirrel monkeys weigh under two pounds.

The important thing is that our brains are about three times larger than would be expected for a primate that weighs as much as we do; and this in itself is significant, for the brain is an extremely power-hungry mechanism that because of its size monopolizes some 20 percent of our entire energy intake, despite its trivial contribution to body weight. What's more, unlike the other bodily organs, the brain demands a constant supply of this large amount of energy (for, interestingly enough, it demands hardly more overall when thinking than when sleeping). There has to be some countervailing advantage for the disproportionate amount of our energy budget that is absorbed by the contents of our skulls; and this advantage can only be sought in our behaviors—and (albeit to a lesser extent) in those of our predecessors, who had smaller yet still larger-than-expected brains. In all fairness I should add here that the apes, and even the monkeys, handsomely exceed (by a factor of at least two) expected values for relative brain size calculated for land mammals in general. For this, of course, they pay the inevitable energetic penalty, while also reaping some rewards in terms of intelligence.

But the matter doesn't rest there, for sheer brain size is far from the full story. The organization—the structure—of our brains is also unique, and it is this that appears to hold the ultimate key to our remarkable cognitive powers. There's a huge amount, of course, that we don't know about how the brain works and especially about how a mass of chemical and electrical signals can give rise to such complex effects as cognition and consciousness. Nonetheless, the way in which vertebrate brains are structured and elaborated tells us a lot about their capabilities.

The brains of all mammals are built on a common plan: something that is, of course, only to be expected in a group of organisms sharing a single common ancestry. It's convenient to view this basic structure as "layered," somewhat like an onion—although the way in which the brain evolved, with new functions and ad hoc expansions being piled onto old structures, makes its organization considerably less tidy than this image implies. The innermost brain components perch atop the spinal cord, at the bottom of the brain, and are those that arose earliest in vertebrate evolution. They are collectively known as the brain stem (or reptilian brain, since the brains of reptiles don't consist of a lot else), and control basic body functions, such as heart rate and breathing, as well as alerting the rest of the brain to incoming signals from the spinal cord. The cerebellum, a posterior expansion from the brain stem, governs balance and movement, and memories related to the control of certain basic learned reponses are also stored there.

With the evolution of the mammals occurred an elaboration of the limbic system (the "smell-brain" of reptiles), a group of structures lying just above the brain stem that mostly function in the control of basic bodily processes and of emotional responses related to survival and reproduction (often known as the four F's: feeding, fighting, fleeing, and sex). The limbic structures also play a role in storing long-term memory, and they comprise the most primitive portion of the part of the brain known as the cerebral cortex. Above them lies a complex area, with a more recent evolutionary origin, known as the neocortex. It is the neocortex that carries out most of what we normally think of as brainwork: cognition, the sophisticated processing of auditory and visual information, mental imagery, and so forth.

In humans, the two hemispheres into which the brain is divided appear to correspond not only to the two sides of the body (the left hemisphere controlling the right side of the body, and vice versa), but at least to some extent to different mental functions. Externally the mammalian cerebrum is wrinkled, most markedly in humans, where a huge surface area of cortex (about fifteen square feet, but less than a tenth of an inch thick) is fitted into the limited confines of the skull. Particularly deep wrinkles define a number of discernible major lobes: frontal, temporal, parietal, and occipital. Traditionally these lobes, or parts of them, have been identified with distinct motor and sensory functions, although recent research indicates that many such functions, as well as more complex cognitive ones, are in fact quite diffusely spread through the cortex.

There are two main points to bear in mind here. The first is that because the brains of higher primates have been produced by an opportunistic process of accretion and elaboration over a vast span of time, they are not tidy pieces of engineering. For one thing, many of the "higher" centers of the brain (that is, those more recently acquired) communicate with each other principally or even exclusively via more ancient "lower" centers. Accordingly, much of the coordination of many "higher" functions is mediated by structures principally devoted to "lower" ones (for example, the four F's). So, however much we may prize our remarkable mental faculties, the old "primitive" brain is always lurking there underneath: one reason, perhaps, that we'll never be the supremely rational beings as which, in moments of hubris, we like to imagine ourselves. It has lately become fashionable to discount the significance of this evolutionary heritage; and it is certainly true that, as we learn more of how information flows between different areas of the brain, the old notion that the accretion of new structures has necessarily led to conflict between sharply distinguished "lower" and "higher" functions has lost much if not all of its force. Still, it is undeniable that our brain does have an evolutionary history. And if the modern human brain is a product of all the multifarious phases of that history, so also must be the behaviors it mediates.

The second point is that the human brain, whatever its marvels, probably does not contain any completely new structures—any structures, indeed, that are not shared with all of our primate—or even mammal—relatives, however humble. Thus we cannot look merely to entirely novel brain components to explain our cognitive powers, however elegant an explanation that would be. What *has* happened over our evolutionary history, however, is that certain parts of the human brain have become enlarged or reduced relative to others and the connections between them modified or enhanced. Even this is not unique to us, though: for while we undeniably have the largest primate cerebral cortexes (about 76 percent of our large brain's total weight), there has been a dramatic increase in the percentage of the brain occupied by the cerebral cortex and supporting structures among higher primates in general. As a proportion of total brain weight, for instance, our cortexes are not hugely greater than those of chimpanzees (which come in at about 72 percent) or of gorillas (68 percent). When we take into account overall brain size, these small differences are, in fact, more or less those that would be expected. However, what is more important in practical terms is that our cortex is much larger in proportion to body size: chimpanzees don't weigh much less than we do and gorillas are often much heavier, but their brains as a whole are only about one-third the size and weight of ours. Further, in apes, the cerebellum makes up a higher percentage of the brain's total volume, making the human cerebral hemispheres relatively larger yet.

Even more significantly, there are differences in the development of the various areas of the cortex among ourselves and the apes. In particular, the association areas—the parts of the cortex that synthesize stimuli from the various sensory pathways and translate them into perceived experience—are greatly elaborated in humans. For example, areas known as the prefrontal association cortex (where a lot of our thinking appears to be done), the temporal lobe, and the inferior parietal region are all better developed in humans than in apes. Several decades ago the great neurobiologist Norman Geschwind noted that the inferior parietal structures lie between the association cortexes for vision, hearing, and somesthesis (broadly, body control); and that, in particular, their posterior part (the angular gyrus), very much larger in humans than in apes, might well serve as the "association area of association areas." Topographically, the angular gyrus is ideally situated to mediate directly between the association areas to which it is adjacent and which are otherwise connected to each other only via the limbic system, a seething mass of nonrational, hormone-mediated urges. Because of this,

Geschwind suggested that the angular gyrus might provide the neural basis of language, by allowing the naming of objects as a result of direct, nonlimbic associations between the centers of vision, hearing, and control of the vocal apparatus.

Well, that was some time ago, and much more has been learned since then about how functions are distributed within the human brain (our knowledge of the apes lags far behind), particularly as new techniques permit the exact localization of activity in the brain while specific mental tasks are performed. Today it is the prefrontal cortex that attracts particular attention as a center of polymodal integration, and it is also beginning to appear that the entorhinal cortex of the temporal lobe—which, as it turns out, is particularly vulnerable to damage in Alzheimer's patients—is crucial in the integration of sensory information, which adds up to what we experience as consciousness. We still have a long way to go, of course; but meanwhile, notions such as Geschwind's point up the undoubted fact that internal brain organization is at least as important as raw size in generating our remarkable mental capacities (Neanderthals, for example, had brains as large as ours, and I shall argue later that they probably did not have language) and that the organization of our greatly expanded association cortex plays a special role in making our unusual abilities possible.

Obviously, if there do exist differences in the development of different brain regions among humans and the apes; and if, as it is reasonable to suppose, the apes in general remain closer to the ancestral condition; and if, again, we can detect any relationship at all between different brain areas and different functions, it becomes natural to ask whether fossil brains can tell us about the sequence of events in human brain—and hence behavioral—evolution. The answer to this one is both yes and no—with, regrettably, a distinct inclination toward the latter. That's why you won't read much about brains when we come to discuss the human fossil record. The good news here is that the cranial vault develops in such a way that it assumes the form the brain itself wants to take on; and since the brain is separated from the bony braincase only by a set of membranes and a few spaces and channels for fluids, the inside of the skull vault preserves a fairly good record of the outside form of the brain. Thus, by making replicas (known as endocasts) of the insides of well-preserved fossil braincases, we can obtain reasonable approximations of the external brain shapes of our extinct precursors—and indeed, in a few cases, nature itself has done this for us. So far, so good.

The bad news, though, is twofold. First, although brain size can be measured quite accurately in endocasts, the details of cortical folding are often rather indistinct. Second, as I've mentioned, new research indicates that many functions are rather diffusely partitioned among several different brain regions, rather than being localized in distinct areas; and we've also recently learned that certain cortical areas traditionally identified with particular functions in fact serve other purposes entirely. Even more importantly, though, the outside surfaces of the cortex give us much less insight than we'd like to have into what is going on *inside* the brain.

Put all this together, and it's evident that it takes a bold scientist to reach any firm conclusions about what the external convolutions of fossil brains actually imply about behavioral capacities. Nonetheless, some brave souls (who call themselves paleoneurologists) have tried—although early enthusiasm has given way to considerable caution. Following various acrimonious disputes over the identification of various landmarks and fissures on endocasts, most paleoneurologists have concluded that it is safer to devote more attention to the overall size and appearance of fossil brains than to analyzing the relative development of specific cortical regions. Unambitious as such scaled-down objectives may appear, however, they are not at all without interest. Brain size increase—the most obvious trend in human evolution and perhaps in the strictest sense the only one—clearly means *something;* and at a general level, documentation of this phenomenon in human evolution is actually not too bad. It is also something that we can correlate quite easily with the behavioral changes we see reflected in the archaeological record—even if, as we'll see, this correlation is far from perfect.

One interesting aspect of gross brain organization that has recently attracted attention and that manifests itself in overall appearance is the matter of cerebral asymmetries. Apes do not show much asymmetry between the left and right sides of the brain, while modern humans do. This is important because, apart from the fact that each hemisphere controls the opposite side of the body, it's believed that in humans the two hemispheres do not perform completely identical functions. For example, in most people the left hemisphere predominates, on average, in the analysis and production of language, in reading and writing, and in tasks that require logic and sequential skills. The right brain, on the other hand, seems usually to function more importantly in musical and artistic abilities, shape recognition (including faces), tasks involving spatial reasoning, and overall perception of pattern. Cerebral dominance is also, of course, the basis of handedness, something unusual outside our own species. According to a recent analysis of human fossil endocasts by Columbia University's Ralph Holloway, it is several million years after the origin of our lineage that we first run into noticeable cerebral asymmetry (suggesting right-handedness), in fossils a little less than two million years old. Interestingly, archaeological studies have indicated that the earliest stone tools were made preponderantly by right-handers. Later humans show further asymmetries, while by the time we reach the recently extinct Neanderthals, we are close to the human pattern of asymmetry (and, it's been claimed, also of external cortical organization).

These observations suggest, if only weakly, that increases in human brain size and organization went to some extent together. And they imply as well that in seeking a starting point for the evolution of human cognition, we can learn a fair bit from the cognitive abilities of the apes.

## CHAPTER 3

# Evolution—For What?

Human beings are the result of the same evolutionary process that produced the entire vast diversity of living things. Yet we cannot help but think of ourselves as somehow significantly "different" from the rest of nature. There's plenty of justification for this, of course; for while our physical peculiarities are no more notable than those of many other organisms, our remarkable cognitive capacities place us in a league to which even our closest living relatives don't belong. As far as we know, for example, we're alone in nature in being able to contemplate our place in it. The story of how we got to our position of apartness is a long one, which I'll recount in the next two chapters. Meanwhile, though, we need to bear in mind that how we view the evolutionary process that gave rise to us profoundly affects how we interpret the evidence of our origins. Hence this chapter, which, in anticipation of discussing the human fossil record, seeks to outline my views of how this process actually works.

Most people who accept that mankind has an evolutionary history tend to think of our evolution as a slow business of perfecting adaptation over the ages—which, if true, imparts in retrospect a certain inevitability to our having become human. Even many paleoanthropologists—students of the human fossil record—find it most comfortable to view our evolutionary history as a long, single-minded slog from benightedness to enlightenment. Paleoanthropologists have even invented a special term—"hominization"—to describe the process of becoming human, thereby reinforcing the impression that there was something unique not just about what we became, but about how we got to be the way we are. This obviously is a dangerous path to follow. For, besides encouraging us in our innate tendency to view ourselves as somehow special, it gives rise to a highly misleading oversimplification of the complex human story. We were not simply propelled to our present place of eminence in nature by some unseen force, natural selection or otherwise; we got here the hard way. In part, science has learned this gradually, as the accumulating facts of our fossil record have forced us to abandon the notion that our biological history has consisted of a simple linear progression. But we have also learned it through an increasing awareness of the complexities of the evolutionary process itself.

## *Natural Selection*

The nineteenth-century English naturalists Charles Darwin and Alfred Russel Wallace are justly celebrated as the fathers of modern evolutionary thought, even though "evolutionary" notions of one kind or another were mooted well before 1858, the year in which they first went public with their ideas. These scientists' great achievement was to hit upon a mechanism—common descent—by which the order evident in nature could be explained. The fact that all living things fall naturally into groups, which in turn form parts of larger groups, is intuitively obvious, and it is reflected in the "folk taxonomies" that are used by all human societies to categorize the living world around them.

This pattern—wherein species belong to larger groups, which in turn form part of larger groups yet—had already been "systematized" over a hundred years before the time of Darwin and Wallace, by the Swedish naturalist Carolus Linnaeus. It was Linnaeus who developed the method of classifying living things into an increasingly inclusive hierarchy of categories (species, genera, families, orders, and kingdoms) that still rules in science today, albeit with many other categories added later. Thus we human beings belong to the species *Homo sapiens* of the genus *Homo.* This, in turn, belongs with certain other genera to the family Hominidae. Similarly, Hominidae and several other families comprise the order Primates; and Primates is but one of the many orders that are included in the kingdom Animalia.

The core of the notion of biological evolution as the explanation of the nested pattern of life-forms lay in the realization that this pattern has resulted from "descent with modification" (to use Darwin's own pithy phrase). Species give rise to other species that do not exactly resemble them. In working out how such modification might be achieved, both Darwin and Wallace were heavily influenced by the views of Thomas Malthus, whose polemical *Essay on the Principle of Population,* published in 1798, had pointed out that the human population had an innate tendency to increase. Other things being equal, Malthus reckoned, human population should double in about twenty-five years. Yet (in Malthus's time and place, at least) it didn't; growth was checked by a shortage of resources that reflected itself in starvation, disease, wars, and a host of other social ills. The weak and the improvident were weeded out; the stronger and the wiser survived. Darwin and Wallace both saw that this scenario applies with equal force to all living things: that in every natural population, far more new individuals are born than can ever survive to reproduce themselves. Wallace calculated that even on quite conservative assumptions, a single pair of birds could potentially produce as many as ten million descendants in a mere fifteen years. Yet the world was not wall-to-wall birds; obviously something was controlling their numbers. And it was this something to which Darwin gave the name natural selection.

The basic observation was simple: All individuals in a population differ from one another, if only slightly, and such differences are usually inherited. Those individuals who are not weeded out of the breeding pool by early mortality or a failure to reproduce will generally be those best adapted to their habitat; and by this process of winnowing, their advantageous heritable features will tend to become commoner with each passing generation, at least so long as the environment continues to favor their adaptations. Variation between individuals will always be there, however; so, should conditions change, selection can (indeed, will) change gear and drive the lineage in a different direction. Natural selection is, then, nothing more than the differential reproductive success of individuals within populations, mediated by the environment. It is a blind, mindless mechanism that lacks any intrinsic direction; but it nonetheless lies at the heart of adaptation and evolutionary change.

But the notion of evolution demands that all life-forms be related by descent from a common ancestor; and while natural selection explains evolutionary change over time, it does not by itself account for the all too evident diversity of the species descended from that ancestor (nobody knows exactly how many living species there are in the world today, but the number is huge: estimates range from thirty million to eighty million–plus). Clearly we need another ingredient in the evolutionary recipe. Here it is. The pattern of diversity in nature resembles a bush with many branches, rather than a single ladder of ascent; so modification within lineages of organisms must be accompanied by the splitting of species into multiple descendant species, just as the main branches of a bush divide into smaller branches and eventually into twigs. But while he acknowledged this, Darwin concentrated on natural selection as the basic component of evolutionary change because to get his ideas accepted, he needed some convincing way to destroy the prevailing notion that species are forever fixed the way God made them. He thus saw evolution as the gradual accretion, under natural selection, of tiny heritable changes over spans of time so vast that they seemed to allow room for even the largest degrees of eventual anatomical divergence. For him, the small-scale changes within populations that we know today as microevolution simply summed up over the aeons to produce the larger-scale discontinuities in nature—between genera, families, orders, and so forth—that we call macroevolution.

### *The Evolutionary Synthesis*

After several decades of confusion, during which innumerable conflicting views of the evolutionary process were aired, Darwin's original view came to be shared by practitioners of the young science of genetics. These scientists came to realize that the large-scale genetic changes that had first grabbed their attention are actually rare phenomena and that genetic changes are typically minor in their effects. And they began to see small innovations of this kind as the wellsprings of the variation upon which natural selection acts. They thus came to view evolution as a matter of generation-to-generation modifications in the "gene pool" of each population as certain genetic variants became commoner at the expense of others, or as new forms of genes were introduced via the process of spontaneous mutation (essentially, the introduction of copying errors into the genetic material).

These developments eventually allowed geneticists, naturalists, paleontologists, and developmental biologists to pool their expertises and arrive at what became known as the evolutionary synthesis. The early tone of the emerging Synthesis was set in 1937 by the naturalist-turned-geneticist Theodosius Dobzhansky, in a book titled *Genetics and the Origin of Species.* Ultimately, Dobzhansky followed Darwin in equating microevolution and macroevolution, seeing all evolutionary phenomena as resulting from gene frequency shifts within lineages, under natural selection. But he did so with some reluctance because as an experimental geneticist, he was able to reproduce only microevolutionary changes in his laboratory, while as an accomplished field naturalist he was acutely aware of the discontinuities that exist in nature. The ornithologist Ernst Mayr followed in 1942 from the systematists' point of view with a book called *Systematics and the Origin of Species.* Mayr pointed out that Darwin hadn't really tackled the problem (the origin of one or more species from another) implicit in the title of his great 1859 book *On the Origin of Species;* and, like Dobzhansky, Mayr was at pains to emphasize the discreteness of species in nature. His especial concern was with the role that the geographical isolation of populations played in the origin of such discreteness. Nonetheless he, too, ultimately conceded that natural selection and shifting gene frequencies lay behind the whole process of evolutionary change.

Finally, the paleontologist George Gaylord Simpson brought his discipline into the fold in 1944 with *Tempo and Mode in Evolution.* Simpson was broadly content to accept the Darwinian view of slow, steady change in the evolutionary histories of the mammals he studied. This view, of course, implied that the fossil record should show an array of gradual transitions between related forms. The problem was that the fossil record exhibited numerous gaps: the expected intermediates between related forms often weren't there. As Darwin had been almost a century before, Simpson was generally willing to see this lack of expected intermediate forms as a reflection of the fossil record's famous incompleteness—for that record is, and always will be, but a shadowy reflection of the diversity of life in the past. Nonetheless, Simpson was deeply concerned by the major gaps evident in the paleontological record: for example, those between the various orders of mammals, members of which embraced anatomies and lifeways as divergent as those of bats, whales, and badgers. Since such differences had clearly emerged rapidly in the very early stages of mammal evolution, Simpson concluded that the appearance of the major mammal groups must have involved unusually fast rates of evolution. To that extent, he regarded the gaps he saw as "real" and sought a special explanation for them; but in the end, he still viewed the problem as one of evolutionary rates rather than of evolutionary modes, with natural selection as the paramount guiding force.

The books by Dobzhansky, Mayr, and Simpson became the defining documents of the evolutionary Synthesis; but although, as we've seen, each of these authors showed a sensitivity toward the complexities of the evolutionary process and to the problem of the origin of discontinuities in nature, in the decade following the end of World War II, formulations of the Synthesis hardened into a simpler dogma. Evolution boiled down to gradual change in lineages due to the adaptive pressure of natural selection. Period. Discontinuities arose simply as a special case of the same continuous process, as geographical accident divided populations and changing environments carried them off on gradually diverging courses, ultimately giving rise to new species.

At this point it's necessary to pause for a moment to consider just what species—the basic "kinds" of living things—happen to be. This is a less straightforward question than it might seem, for the precise nature of species has been energetically debated just about as far back as the written record goes; and if there is one thing certain in science, it is that this debate will, in some form, continue indefinitely. Nonetheless, most zoologists would agree nowadays that the most essential characteristic of species is their reproductive cohesion: species are regarded as the largest populations within which fully effective interbreeding among individuals is possible. To put it the other way around, the boundaries between species are defined by reproductive incompatibility—not by the anatomical differences that are all that paleontologists have to go on. Further, it's not at all clear that the origin of reproductive incompatibility (which itself turns out to be an extraordinarily difficult quality to define precisely or even in some cases to recognize) has anything necessarily to do with the accumulation of morphological novelties. Yet the Synthesis reduced the process of speciation—the production of new species—to nothing more than a passive result of adaptation over long stretches of time. And the simple repetition of such speciation by divergence was seen as the mechanism that ultimately gave rise to new genera, new families, new orders, and on and on. Finally, this scenario more or less directly associated the passage of time with morphological change within lineages. As it turns out, all of these propositions are debatable at the very least; but, as the Synthesis congealed, the problems implicit in it went largely ignored.

One reason for the success of the Synthesis is that the gradualist metaphor for speciation is beguiling in its simplicity; and it was particularly attractive to the geneticists, whom it firmly established as the guardians of evolution's mysteries. For speciation, perhaps the most significant natural process of all, was reduced to a matter of genetic change in populations under the guiding hand of natural selection. This view turned out to be also quite congenial to systematists, the scientists concerned with the diversity of the living fauna and flora. For after all, even though they had to concede that species had no objective reality in time, systematists could still regard them as discrete entities as they carried on with their daily business of sorting out the diversity of living organisms. It was, indeed, only the paleontologists who were shortchanged; but they were shorted with a vengeance. For time, the dimension that was uniquely theirs, robbed paleontologists of any practically useful concept of the species, their basic unit of study.

This is because, according to the Synthesis, species inexorably evolve themselves into other species by tiny incremental changes. Transformations from one species to another are thus accomplished with no boundary between parent and daughter that a paleontologist could hope to detect: indeed, with no boundary at all. Of course, change within lineages there is, and in abundance. But the expectation was that known representatives of each lineage had been linked by a smoothly intergrading series of ancestors and descendants. Only an arbitrary line could be drawn between successive species: a line that had inevitably to separate a parent from its own offspring. This should have made students of the fossil record uneasy; but in fact the period following World War II witnessed the astonishing spectacle of paleontologists congratulating themselves on the deficiencies of their data: gaps in the fossil record, periods from which no lineage members were known, were hailed as convenient places to recognize species breaks without having to confront the unpalatable process of separating parent and offspring.

Species thus became segments of lineages defined for practical purposes by random discontinuities in the geological record. They were seen as ephemeral and thus somehow not "real." Yet to know to which species an individual fossil belongs is the single most important question you can ask of it; and it is a biological question, not the luck of the geological draw.

To say this is not to belittle in any way the enormous achievements of those who contributed to the Synthesis. This new view of evolution, although perhaps incomplete, had the merit of sweeping away a vast amount of intellectual debris. Gone forever were a host of myths and misconceptions, ranging from views of species as idealized "types," through notions of innate impulses toward change within lineages, to the idea that macroevolution is achieved through major genetic "jumps." Nobody could ever again look at the evolutionary process without very consciously standing on the edifice of the Synthesis. And this edifice was not only one of magnificent elegance and persuasiveness; it had also brought together practitioners of all of the major branches of organismic biology, ending decades of infighting, mutual incomprehension, and wasted energies.

## *The Synthesis and Paleoanthropology*

Throughout all the ferment of rethinking that gave birth to the Synthesis, paleoanthropologists stuck very much to the sidelines. Indeed, for most of the time, they seem barely to have noticed what was going on. This is less surprising than it might appear, for, at least partly because of the belief—at times unconscious, at others more explicit—that human evolution is somehow "special," paleoanthropology has always stood apart from the mainstream of evolutionary thinking. It has occasionally taken in evolutionary and systematic concepts from other areas of biology, but it has exported few if any. What's more, even when paleoanthropology has embraced new developments in evolutionary theory, it has generally been very slow to do so—with significant consequences.

As I've already mentioned, in the years between its founding and its eventual triumph in the 1950s, the Synthesis gradually took on a simpler and more unyielding aspect. Freed from the need to fend off opponents' objections, its adherents were eventually able to offer a tidier vision than before of the evolutionary process; and it was in this purer, more fundamentalist form (evolution proceeds by gradual modification of gene frequencies in lineages, guided by natural selection; macroevolution and microevolution are the same thing) that the Synthesis was eventually absorbed by paleoanthropology.

Equally significant is that when the bastion of paleoanthropology finally fell to the Synthesis, it tumbled with a vengeance. Insulated during the 1930s and 1940s from mainstream developments in evolutionary theory, students of the human fossil record came ultimately to embrace the fundamentalist version of the Synthesis with the zeal of converts. In large part this was because the Synthesis happened to be entirely congenial to their traditional way of doing things. Most paleontologists seek to uncover and explain the evolutionary histories of diverse groups of organisms; they are looking for the origins of arrays of species. In stark contrast, paleoanthropologists have seen their job essentially as one of tracing the origins of the single species *Homo sapiens* as far back into the remote past as possible. Although a lingering typological tradition allowed for a multiplication of named species in the human fossil record, the attempt to project our human lineage back into the past both produced and reflected an essentially linear mind-set; and in the Synthesis, this mind-set found the perfect theoretical framework. Indeed, so ideal was this pairing of preoccupation with theory that it enabled the founders of the Synthesis to exert an influence on the paleoanthropological outlook that was as direct as it was powerful.

For paleoanthropologists didn't just passively absorb the new evolutionary principles. They also listened, enthralled, as Dobzhansky, Mayr, and Simpson told them how to interpret the fossil record that was their domain. The fact that none of this triumvirate, save perhaps Simpson (the least intrusive of the three), can have had more than a passing acquaintance with the human fossil record seems to have counted for little in the face of these gentlemen's formidable reputations and the authoritative stances they assumed.

Dobzhansky began to share his insights with paleoanthropologists as early as 1944. Pointing out that all populations are variable and that most species consist of mosaics of distinctive local populations, he concluded that while human evolution since Java man (a millionish-year-old form of human now allocated to the species *Homo erectus*) had been pretty eventful, all of the events concerned had taken place within the confines of the single species *Homo sapiens.* Further, over the entire span of human evolution, Dobzhansky declared, "as far as [is] known no more than one hominid species existed at any one time level."

Six years later, Mayr weighed in with a survey that went further back into the past but which similarly reached the conclusion that no more than one species could be discerned in the human fossil record at any point in time. What's more, said Mayr, in the entire span separating modern humans from the first bipeds then known (the primitive and small-brained South African australopiths), he could find evidence for only three species: *Homo transvaalensis* (the australopiths), *Homo erectus*, and *Homo sapiens.* Only three (vastly different) species in what we now know to be three million years that witnessed immense physical change in our lineage!

There's no doubt, though, that in propounding these views, Dobzhansky and Mayr performed a very considerable service for paleoanthropology, a science in which name inflation had been rampant. An essentially nonbiological view of species and a failure to recognize the importance of within-species variation among individuals and populations had allowed far too many species names to creep into the literature of human evolution, almost as if each new fossil needed its own formal name to validate it.

By reducing the plethora of names on offer to a very few, Dobzhansky and Mayr had unquestionably performed an essential housecleaning. But in hindsight there's equally no doubt that they went much too far, creating the illusion of simplicity in a picture that is actually far more complex. As part of an ongoing scientific enterprise, this was no bad thing; after all, other badly needed reappraisals of diversity in the primate fossil record have been spurred initially by the excessive pruning of species. But, regrettably, Dobzhansky and Mayr had, unconsciously or otherwise, also resuscitated the powerful notion that human evolution had consisted simply of the long, undeviating slog from primitiveness to perfection about which I wrote earlier in this chapter: a notion that was strongly underpinned by the persuasive but incomplete body of theory that they had themselves created. Superficially supported by the observation that brain-size increase with the passage of time has been a fairly consistent signal in the human fossil record, this distorted view of how we got to be the way we are has been a pervasive theme in paleoanthropology throughout the second half of the twentieth century and shows no sign of fading yet.

## *Equilibrium Punctuated*

Given that paleontologists had been relegated by the Synthesis to the humble clerical function of documenting the results of a process to which others held the key, it is hardly surprising that complaints that theory and observations didn't match first began to be heard from the direction of paleontology. For years paleontologists had been dimly aware that fossil species often appeared to be "real" in a way that the Synthesis didn't allow. Instead of gradually transforming one into another over the aeons, new species tended to appear suddenly in the geological record and to linger for varying periods of time. When their time was up, they would vanish as abruptly as they had arrived, to be replaced by other species that might or might not be their close relatives. As far back as Darwin, the fossil record's failure to agree with the expectation of gradual change had been explained by its famous incompleteness. And it is, of course, quite true that this record preserves only the tiniest fraction of one percent of all the individuals who have ever existed. Nonetheless, it is equally true that while the number of fossils known has expanded enormously since Darwin's day, the basic pattern of discontinuities that the fossil record reveals has not changed at all. Eventually, the diaphanous nature of the emperor's clothing was bound to be noticed.

Ironically, since it had also been the home of Mayr and Simpson—and a near neighbor of Dobzhansky—during the formulation of the Synthesis, the first intimations that this theory might be less than complete came from New York's American Museum of Natural History. In 1971 my American Museum colleague Niles Eldredge published the results of a study of a group of trilobites, ancient sea-bottom-dwelling invertebrates. In the trilobites of interest to him, whose fossils occurred in rocks of both New York State and the Midwest, Eldredge found a remarkable *lack* of evolutionary change. In the Midwest, for example, only a single significant change marked a period of six (then he thought eight) million years. Trilobites have compound eyes with multiple rows of lenses; and during the first part of this period, those studied by Eldredge had eighteen such rows. Suddenly, however, these creatures gave way to otherwise identical trilobites with only seventeen rows. The New York pattern was similar, except that a single transitional locality, representing a geological instant, yielded fossils of both kinds. What's more, this site was millions of years earlier in time than the transition in the Midwest. To Eldredge, who was not prepared to fall back on the inadequacies of the fossil record to explain it, this observation could mean only one thing: that a speciation event (the origination of a new species) had occurred in what is now New York (and had been caught in the act, as it were, in that single New York locality), while in the Midwest things had stayed as they were for millions of years until an environmental or geographic change had allowed the new species to sweep in from the East and replace the old one.

The pattern, then, was of a long period of business as usual for the older species, followed by rapid local extinction as a new species moved in from elsewhere. If the evidence of the New York record was to be believed, the origin of the new species had had two characteristics: it had been short term, and it had not involved a great deal of morphological change, at least in characteristics observable in the fossils. To explain this species change, Eldredge invoked the allopatric ("other country") speciation concept: a notion that, ironically, had been most fully explored by Ernst Mayr. The basic idea here is that new species arise from old when formerly interbreeding populations are physically divided by a barrier such as a mountain range or a seaway. Unable to exchange genes, the two "daughter" populations will begin to diverge genetically as mutations and other genetic changes accrue. Because small gene pools are inherently more unstable than larger ones, Mayr felt that events leading to genetic incompatibility—speciation—were most likely to occur in small isolated populations peripheral to the parent population, rather than in the parent itself. Most importantly, the acquisition of genetic incompatibility by this mechanism is a short-term process compared to the vast spans of time involved in the Darwinian vision. And it seems, at best, to be only indirectly related to morphological change.

Long periods of species stability interrupted by brief events of speciation, extinction, and replacement: a very different vision of life history from that of the Synthesis. True, the framers of the Synthesis had scrupulously acknowledged that rates of evolutionary change could vary enormously. But they had ascribed such variation simply to fluctuations in the strength of natural selection. For them, evolutionary change was an essentially continuous process that would promote better adaptation to static environments when it was not producing accommodation to shifting ones. Eldredge, in contrast, was proposing that in the case of his trilobites, speciation had been the basis of evolutionary change, rather than its passive result. The dominant signal he discerned in the fossil record he had so closely examined was one of stability, not of gradual change. And with thousands of specimens to work with, he wasn't prepared to explain away what he saw as the product of an inadequate record. Eldredge may not have fully realized it at the time, but he had set the stage for paleontologists of all stripes to go back to their fossil data and to rethink their views of the evolutionary process.

Obscure papers in technical journals rarely spur such rethinking; but in 1972 Eldredge joined forces with the land snail expert Stephen Jay Gould, who had noticed a similar irregularity of change in his own subjects of study, to publish a paper provocatively entitled "Punctuated Equilibria: An Alternative to Phyletic Gradualism." This contribution, both more emphatic and more general than Eldredge's earlier paper, attracted immediate attention. Noting that preconceived ideas about the evolutionary process affect the ways in which paleontologists view their fossils, Eldredge and Gould proposed that perhaps the famous "gaps" in the fossil record were actually telling us something after all. Maybe these discontinuities were not simply artifacts of incompleteness, but instead reflected an underlying reality. In which case, might it not be unwise to expect to see steady change through time in any local rock record? After all, if allopatric speciation is the key to innovation, new species will almost invariably have arisen not only suddenly, but someplace other than where you happen to be looking.

Eldredge and Gould were, in fact, proposing that evolution is not a gradual process, but rather one that proceeds by fits and starts ("evolution by jerks," as an unkind early critic put it). Species, far from being nothing more than arbitrarily defined segments of lineages, are real entities. Like individuals, they have births (at speciation), life spans (of greatly varying lengths, but unmarked by significant change), and deaths (at extinction). What's more, on the wider evolutionary stage, species play a role similar to the one envisaged for individuals in traditional Darwinian theory. Species can become extinct for many reasons, including being outcompeted by other species, among them their own descendants: in other words, just as individuals vary in their ability to compete with each other, so do species. Trends in the fossil record—such as human brain-size increase with time—are often cited as evidence for gradual, directional change; but it is in fact much more probable that trends result from the winnowing of species by competition among them, than from directional selection within lineages.

Such a radical departure from received wisdom was bound to cause a stir, and many paleontologists initially objected to Eldredge and Gould's conclusion that their mechanism of "punctuated equilibria" accorded much better with the patterns observed in the fossil record than did traditional "phyletic gradualism." Two decades later some still do object; but—except in paleoanthropology—they are in a diminishing minority. In the early days, though, fault was also found with the theory itself. Many claimed, for example, that punctuated equilibria was merely the discredited notion of "saltationism" in a new guise; but this was to confuse speciation (already squeezed into the Synthesis by Mayr and others) with the major leaps to which nobody, least of all Eldredge and Gould, still subscribed. Others alleged that the authors of punctuated equilibria were hostile to the notion of adaptation, and this was equally untrue: organisms may not be as exquisitely fine-tuned to their environments as the Synthesis suggested, but adaptation is a notion that is fully incorporated into the Eldredge-Gould view.

Of course, it was inevitable that in their zeal to convince colleagues of their viewpoint, Eldredge and Gould should overstate their case in a few respects. For example, in my view it was putting it a little strongly to insist that all morphological innovation takes place at or around speciation, that no significant change accumulates within the lifespan of species, or that speciation itself is a supremely rare and difficult event. Refinement of the notion of punctuated equilibria thus continues, not least by its founders; but by extending the concepts of the Synthesis, this view has provided a more complete account of the evolutionary process than any we have had before—one that has greatly illuminated our understanding of the actual fossil record of evolution.

## *Group Selection and Immortal Genes*

Macroevolution's demotion by the Synthesis to nothing more than a special case of microevolution had a major effect on postwar evolutionary biology. Investigation became centered largely upon the origin and development of adaptations within lineages, and the notion took root that natural selection inevitably acted to "optimize" adaptation. The classic example of the latter was the number of eggs per clutch a bird might lay: too few would reduce the parents' representation in the next generation, while too many would handicap the parents' ability to ensure the survival of each offspring. Necessarily, there had to be an optimal number of eggs—even though this number might be made nonsense of by the behavior of certain individual bird species. The ground hornbill of southern Africa, for example, produces two eggs per clutch. But the second egg is laid a week after the first, and by the time the younger chick hatches, the older sib is strong enough to dispose of it when the parents are not looking. This behavior certainly benefits the older offspring, who no longer has to compete for the parents' attention; but how it squares with the optimal use of the parents' energies is far from apparent.

Nonetheless, the notion of optimization was still an attractive one as long as you simply looked at supposed adaptations one at a time. But as the term came to be applied to more or less any structure to which an investigator thought that he or she could assign a function, every individual came to contain a whole host of adaptations. The obvious difficulty this raised, that reproductive success could only vote up or down on the overall performance of an organism—not on that of its individual components—tended to get lost as attention became increasingly focused on within-lineage transformations in particular characteristics. One school of thought went further still, viewing individual genes (many of which are involved in most adaptations, and most of which are involved in more than one) as the unit of natural selection.

Among the most influential of the adaptationists was the biologist George Williams, who in 1966 published a book aimed at countering the suggestion that natural selection might somehow work to influence not just individuals, but entire groups of organisms (a notion known as group selection). Selection, Williams maintained, could act only in situations where at any point in time there existed variations contributing to the relative reproductive success of individuals. Such situations were invariably found *within* species, so there was no way in which selection could be "for the good of the species" as a whole. Perhaps most significantly in the light of later developments, Williams threw his support behind the idea that adaptations are principally devices to facilitate the survival of the underlying genes. This theme was hugely popularized a decade later by Richard Dawkins in his best-selling book *The Selfish Gene.* There, Dawkins argued that selection acts not so much on the physical characteristics of organisms as on the "immortal" genes themselves. Individual organisms are little more than vehicles for the genes, which compete through them for representation in the next generation. Anything bigger than the gene, Dawkins declared, is too "large and too temporary a genetic unit to qualify as a significant unit of natural selection."

This viewpoint more or less eliminates anything larger than the individual gene as an actor of any kind in the evolutionary process; and it is made possible only by ignoring all of the economic aspects of existence (including that of the genes themselves, which have to coexist with others in a highly complex interaction). It also neglects the fact that genes merely represent the lowest level in a hierarchy of genealogical complexity that continues upward through individuals and local populations all the way to entire species. In this hierarchy, each level has a distinctive and essential evolutionary role to play. What's more, the highly reductionist view of the selfish gene runs into the problem already noted, that for all the many millions of genes a complex organism contains, selection can only vote on the success of the individual as a whole.

Reductionist ideas are always attractive to the human mind, however, and the notion of the selfish gene was soon integrated into the newly hatched concept of kin selection. This idea was devised in the 1970s to explain how apparent acts of altruism (which harm the actor, or at least do not benefit him or her directly) may actually assist him or her (or at least their genes) by propagating those genes. This will happen when the beneficiaries of the actor's altruism are relatives, with closely similar genotypes. The actor expends energy (or even his life), without reaping any economic benefit, but the chances of propagation of much of the information coded in his genes will be enhanced. Of course, kin selection is not routinely observable in all social systems; but its marriage to the selfish gene concept nonetheless gave birth to the cult of sociobiology. Here the argument is that within any social system, individual organisms tend to show some degree of cooperation (and sometimes of altruism); and because of the widening network of relatedness within interbreeding groups, kin selection will tend to reinforce cooperation and sociality. To the extent that behaviors are genetically determined, they will be at least to some degree under the influence of natural selection. Which, ironically, brings us back to a form of group selection, since a tenet of sociobiology is that the behavior of individuals in cooperative societies influences the success of those societies as a whole. Here the individual is not seen as a single economic entity, but as a member of a group upon whose overall success his or her survival depends.

There can be little doubt that such feedback has been a powerful influence in molding both individual behaviors and social structures in the highly, if mindlessly, organized insect societies that first stimulated sociobiological inquiries. But its relevance to primate (and particularly human) societies is much more questionable, and the Harvard entomologist E. O. Wilson's claims for such relevance caused an immediate outcry when he published *Sociobiology: The New Synthesis* in 1975. Nonetheless, even though nobody would claim that normal human behaviors are under anything like the same kind of control as those of ants, the interference of genes in human behavior has become a perennial theme that refuses to go away. It has surfaced most recently under the guise of evolutionary psychology, an attractive but profoundly flawed approach to explaining human behavior that has garnered widespread media interest in recent years.

We'll return to evolutionary psychology and similarly reductionist behavioral gambits in chapter 6.

## *Evolutionary Theory Today*

The details of the evolutionary process are as hotly debated today as ever, and it would be pointless to try to represent all sides of this multifaceted argument here. Nonetheless, since even subtle differences in the ways people think evolution works can have huge effects on how they interpret the fossil record, I should at least declare my own hand as a working paleontologist.

In my view (and I'm not alone), the key to understanding the evolutionary process lies in acknowledging not only the existence but also the significance of the hierarchy of biological organization to which I've already referred. This is the hierarchy that starts with the genes and proceeds up through the individual to local populations, species, and maybe even beyond. All of these levels are involved in the evolutionary process, but each participates in its own way. Mutations of the genes and recombinations among them provide the underlying variation on which natural selection capitalizes. Through promoting the reproductive success or failure of individuals with more or less favorable inherited traits, natural selection fosters local population adaptation to specific environments. Virtually by definition, local populations are the only ones likely to have a relatively homogeneous habitat—or set of habitats—to which adaptation is possible; and because of this, it is local populations that are the true dynamos of evolutionary innovation. New traits originate with individuals, of course; but it is only the establishment of such traits as norms in the population that constitutes innovation in a true evolutionary sense.

In short, then, individuals don't evolve; populations do. But which populations? In one sense, of course, it is the reproductively cohesive species that constitutes the definitive population. But in another—and equally important—sense, species are essentially abstractions. For, as just noted, they usually consist of multiple local populations that occupy a variety of environments—or at least of microenvironments. The critical consequence of this is that evolutionary tendencies within species will not tend to be unidirectional. Rather, such tendencies will be toward divergence, as each local population within the species plows its own adaptive furrow in its own local habitat. And we know for a fact that local populations do routinely develop their own heritable peculiarities, even though it's possible to argue—though I don't recommend it—that in most cases, natural selection is something that we have to assume, rather than something we can know. Only extremely recently, for example, has any direct experimental evidence at all become available to confirm that natural selection does indeed function within populations.

Local populations thus almost always become distinctive in some way; but it is the event of speciation that "fixes" their innovations and gives them historical validity. For while distinctive local populations may indeed have their own economic existences, they are still linked to the rest of their species by reproductive continuity, or at least by its potential. Habitats are constantly fluctuating, and the isolated or quasi-isolated populations within which innovations arise are always at risk of finding themselves and their genetic peculiarities reabsorbed into the parent population. This is where the isolating mechanisms that assure at least some degree of reproductive incompatibility between related populations come into their own. By disrupting reproductive continuity, they create new distinctive historical entities out of ephemeral units that were previously at risk from reabsorption or from local extinction. In the longer run, the new species may not survive; but its existence, however brief, will at least have been objective and independent of any other. And if, conversely, it is successful and enlarges its range, it will take its distinctive characteristics into new habitats, where the twin processes of local innovation and interspecies competition will start all over again. Thus the species entity acts principally to define the bounds within which innovations may accumulate, while the speciation process singles out new actors for the evolutionary stage.

Two further points are important here. One is that isolating mechanisms—which appear to require a restructuring of the population gene pool in a way that incorporating new physical variations does not—are most likely to become established in small and at least quasi-isolated populations. This is because the gene pools of small populations are inherently far less stable than are those of large ones, since in the latter, any innovations that arise will tend to be swamped by the inertia of established genotypes. The other point, which I've already briefly raised, is that while small gene pools will more quickly incorporate favorable novelties than large ones are able to, the acquisition of isolating mechanisms is not necessarily the same thing as the acquisition of new adaptations. Indeed, the two may be totally unrelated. This is why in some instances we can observe the accumulation of enormous interpopulation differences within a species without any hint of genetic isolation, while in other cases closely related but reproductively distinct species may appear virtually identical to the human eye. Further, although there's an awful lot that we still have to learn about the mysterious process of speciation, it is already abundantly clear that all morphological or behavioral innovations, adaptive or otherwise, have to arise *within* freely reproducing populations. However tempting it may be, we cannot routinely use the arrival of new species to "explain" the appearance of new morphologies in the fossil record—although as a result of the record's imperfections, the two may show up together in practice.

It's also important to understand that not all morphological innovations that become established within populations necessarily represent adaptations in a strict sense. There are plenty of chance factors that can intervene in the acquisition of new morphologies, especially in small populations. This doesn't, of course, mean that an important character that has become randomly fixed is forever exempt from selection of some kind. Not only may local circumstances change and make such characters "useful," but once genetic isolation has been achieved between a local population and its parent species, the two now-independent species are free to compete should contact ever be reestablished between them. This shifts the winnowing effect to another level, one that embraces all population attributes. And, as Eldredge and Gould suggested, we should look to the differential survival of competing species to explain the long-term trends that we see in the fossil record.

Yet the differential survival of species is not always a question of excellence of adaptation, either. Environments may be relatively stable or they may not; but when they change, they usually do so pretty rapidly, at rates with which adaptation by natural selection would be hard put to keep up. When such change occurs, the quality of your adaptation to your old habitat is irrelevant, and any competitive advantage you might have had may be eliminated at a stroke. If the external change—which will usually involve the uncomfortable arrival of new species, including perhaps close relatives—is great enough, you are left with only two viable alternatives: migration to a more congenial habitat or extinction. In extreme cases the latter is virtually the only course open, as periodic mass extinctions over Earth's history have dramatically demonstrated. But whatever the case, adaptive change on the spot seems rarely if ever to be an option. Environmental shifts generally happen too quickly to be tracked immediately by natural selection. They also have the effect of allowing the rapid colonization of tracts of territory by new arrays of species, leading to competition and to rapid faunal turnover. As we'll see, our own genus *Homo* may have emerged in association with an environmentally driven "pulse" of this kind.

Genes, individuals, populations, and species are thus all winnowed in one way or another over evolutionary time, and it is the results of this winnowing that we see reflected in the fossil record. Given this, what kind of pattern should we expect our fossils to show? Eldredge and Gould rejected the notion that we should look for evidence of slow generation-by-generation change. Instead, they saw speciation as a rare and difficult process, resulting in the occasional short-term replacement of stable lineages. But my own observations of extant primates suggest that very closely related species, while genetically isolated from each other, usually differ rather little in the hard tissues, which are all that preserve in fossil form—however dramatically different their living representatives may appear to the eye in such features as coat color and hair distribution. This is hardly surprising, of course, when you consider that any difference that sets off a species from its closest relative must have arisen *within* the common parent species. But it does strongly imply that speciation must be a pretty frequent occurrence to have given rise to the degrees of differentiation we actually see.

What speciation achieves, then, is the shifting of what one might call the genetic and morphological centers of gravity of the parent and daughter species. On average differing at least somewhat from each other before speciation, each species is now free to accumulate more variation and hence more potential species difference. Their descendant lineages will thus ratchet away from each other in a variety of directions, and the particular evolutionary trajectories they describe will result from the relative success of the species that comprise each lineage, whether or not they ever find themselves in direct competition. These species will probably vary in time in much the same way they do in space, hence within fairly strict limits. They will thus more or less conform to Eldredge and Gould's criterion of stability, although in practice it may turn out to be difficult or impossible to discriminate members of closely related species from each other reliably on the basis of fossil remains of bones and teeth.

The implications of all of this for patterns in our own human fossil record are pretty obvious. The morphological differences between us and our earliest known ancestors are profound. If we use the average amount of morphological difference between closely related primate species as some kind of yardstick (obviously, only a very approximate one), it seems evident that a large number of speciations—far more than the number of species conventionally recognized in the human fossil record—must have been required to achieve the remarkable degree of morphological shift we observe over our biological history. This is so even though recent findings in molecular biology (which now, for the first time, is really beginning to tie genetic mechanisms in with those of evolution) suggest that quite minor genetic changes may sometimes have far-reaching developmental and morphological effects. And although I'm prepared to discern some evidence that short-term radical shifts of this kind have been significant at certain points in the evolutionary history of our family, my feeling is that we have almost certainly underestimated the number of species actually present in the known human fossil record.

As a result, we have tended to view the history of our kind as less eventful than it actually was. At the same time, we have made our own species appear more central to the story of our family's evolution than is appropriate for what is in effect just one more terminal twig on a large branching bush—though admittedly the only surviving one. For, as I've noted, from the very earliest days of their profession, paleoanthropologists have been preoccupied with the task of projecting the origins of the single species *Homo sapiens* as far back into the past as possible—almost invariably in a linear fashion. And although paleoanthropologists are certainly unusual in having only a single living species of interest—most paleontologists are concerned with the origins of diversity among large groups of species—it's clear that this traditional effort has been profoundly misguided.

The acceptance of an intricate pattern of speciations and extinctions in the biological past of our family accords well with what we know about the complex environmental changes that beset Earth during the Ice Ages, roughly the period during which our genus has been on the planet. Such acceptance also obliges us to rid ourselves, for once and for all, of the persistent notion that we are the final result, whether perfected or not, of a steady process of improvement. It's also important for us to acknowledge very explicitly that, remarkable as we humans unquestionably are, we did not come by our special features as the result of a special process. We came by them honestly: through the workings of exactly the same mechanisms—poorly understood though some of them may be—that gave rise to all other living things.

## *A Final Caveat*

In this chapter I have tried to emphasize the importance of populations and species in the evolutionary process and to spell out their roles, so far as they may be understood, in the production of evolutionary trends (which, we must remember, exist only in hindsight). It is critical not to lose this population-centered perspective; yet, in practice, most discussion of the evolutionary histories of groups—as distinct from the theoretical exegesis of evolutionary processes—has been dominated not by the delineation of the histories of species, but by consideration of the "evolution" of particular physical characteristics or functional complexes. In the case of our own lineage, for example, paleoanthropologists have been most often concerned with issues such as the evolution of bipedalism, or of brain size, or of details of the skull and dentition. Yet it is evident not only that innovations must arise within species, but that each species necessarily consists of aggregations of individuals, whose personal and corporate survival depends on their smooth functioning as entire organisms.

In other words, as fascinating and important as they may be, characteristics and functional complexes cannot and do not exist in isolation. Neither natural selection nor random chance nor larger-scale sorting can act upon one bodily system alone, whether within or across species. As I noted earlier, nature can only vote up or down on the entire functioning organism; it cannot favor or disfavor variants of one trait or organ system without affecting other such features. Of equal importance, it is species, in all their individual complexity, that are the winners or losers in the evolutionary game. Hence it is, for example, totally fruitless in a fundamental sense to debate—as is routinely done—whether upright walking originated as a locomotor adaptation, or as a thermoregulatory mechanism, or as a means of increasing visual range, or as a way of discouraging predators whose interest is triggered by horizontal silhouettes. The plain fact is that this behavior simply arose in the first bipedal species in our lineage. Once upright walking was adopted (probably, it increasingly appears, as an adjunct to, rather than a total replacement of, climbing abilities), members of the species concerned were necessarily affected by all of the factors I've mentioned, and by many more. Those early bipedal individuals—and species—were functioning wholes, and they were necessarily successful as wholes, not merely as vehicles for one or another of their "traits" or "adaptive complexes." Similarly, the evolution of brain size in hominids sums out not as a simple long-term increase in brain volume, but as the aggregate history of a succession of species that showed some brain enlargement and elaboration (and presumed increased behavioral complexity) compared to others. The preferential survival of certain larger-brained species must necessarily have taken place within the context of all of the other attributes that contributed to—or militated against—their relative success.

To return, then, to the question posed in the title of this chapter, evolution is "for" nothing whatsoever. As a process, it is, as I've said, famously "blind." Most importantly, anatomical and behavioral novelties *arise* for no purpose; they are simply thrown up by a variety of genetic processes that in themselves have nothing to do with evolution. Of course, subsequent winnowing via the mechanisms I've discussed may certainly favor the spread of some feature or other. But as we follow the events of human evolution in the next two chapters, let's not lose sight—as it's so easy to do—of the crucial role that is played in this story by larger units than genes and character complexes, or even individuals.

One last word on terminology. The notion that organisms are adapted to their environments long precedes Darwin; it was, indeed, regularly used well before his time in support of the excellence of God's design, for all adaptations were *for* something. With the introduction of evolutionary thought, the term *adaptation* has not surprisingly acquired a number of different shades of meaning. As we have seen, characters do not initially arise *for* anything; but once in existence, they can be exploited as required. Strictly speaking, an adaptation is a character whose function can be identified and that appears to have fulfilled that function since its origin. Some characteristics, though, arise and function in one context before being exploited in another. Among birds, for example, feathers may have emerged first of all as a device for retaining body heat, only later being co-opted as part of the flight mechanism. Originally, then, feathers were adaptations for improved thermoregulation but were not adaptations for flight. Instead, they slumbered as latent "preadaptations" for flight, which the bird ancestor was able to capitalize upon once the occasion demanded it.

The case of feathers illustrates an extremely common pattern in evolutionary history, one that is underwritten by the opportunistic character of natural selection. A recent attempt to clarify the terminology of adaptation, motivated by the fact that the term *preadaptation* had over the years acquired some dubious overtones, has suggested the following: All useful features of organisms should be called *aptations. Adaptations* are only those features that were fixed in the context in which they are now employed. Features that originally arose in one context but were later co-opted for use in another are *exaptations.* Human hands, for example, turn out to be wonderful devices for playing the piano; but that's hardly why we acquired them. Reluctant as one is to see yet more jargon creep into the evolutionary lexicon, the distinction here is actually a very important one; for, as we shall see, exaptations have played a significant role in human evolution.

CHAPTER 4

# *Starting Out*

However you look at it, humanity's biological history goes back a long way: in one sense, all the way back to the first microscopic life-forms that arose over 3.5 billion years ago. But when we search for the origin of those features that make *Homo sapiens* the unique species it is, we don't have to look too far back into the past. The human story begins with the emergence of our family, Hominidae; and the first hominids appeared not much more than five million years ago, in Africa. People are very often surprised to learn that, genetically, we are well over 98 percent identical to chimpanzees; but placed in the perspective that we have lived some 99.9999997 percent of our entire evolutionary history in common with the African apes (and, as I've heard it alleged, that we share 40 percent of our genes with a banana), this figure seems much less extraordinary. Still, less than 2 percent makes a remarkable difference; and in the remaining chapters of this book we will explore this difference and how it emerged.

## *In the Beginning*

The first known hominids are recently discovered handfuls of fragments from Ethiopia and Kenya that date in the range of about 4.4 to 3.9 million years (myr) ago. *Ardipithecus ramidus*, first announced in 1994, consists of some skull and limb bone fragments, plus a few teeth (and apparently now also most of a fragmented skeleton, though this has not yet been formally announced, let alone described) from the 4.4-myr-old Ethiopian site of Aramis. In view of the great antiquity of this new form, its discoverers predictably enough claimed that they held in their hands the "root" species of Hominidae: the founder of our lineage. Others weren't so enthusiastic. They pointed among other things to the narrowness of the molar teeth and their thinnish enamel coating: characteristics that are disconcertingly apelike. Still, if the apes are our closest living relatives, mightn't a very early hominid—the best part of a million years older than the next oldest hominids—be expected to look like this? There was certainly room for speculation. One possible feature that indirectly argued for hominid status was found in tiny braincase fragments that, though "strikingly chimpanzee-like," were said to indicate that their possessor might have had a rather forwardly shifted foramen magnum (the hole through which the spinal cord exits the skull). If this was indeed the case, it is highly probable that the creature had habitually held itself erect.

Which, if true, is important. Paleoanthropologists have argued loud and long over the adaptations of the first hominids. In the early years of our century, when there weren't a lot of hominid fossils around to compare, many were so impressed by the large size of the modern human brain that they believed it was sheer brainpower that had dragged humans upward from the ape masses and that had set us on the road to our present eminence. Accumulating fossil evidence quite rapidly made this viewpoint untenable, and although various alternatives have been entertained from time to time, it's been evident for several decades now that the adoption of a form of upright bipedal walking was the crucial behavioral/anatomical shift that got our lineage started. That shift is a particularly attractive candidate for this role because it brought along with it a whole series of consequences, including the freeing of the hands from the demands of locomotion. What's more, if postural change was not the key to our lineage's origin, we have precious few alternatives in reserve, for it has lately become fashionable to refer to our more remote known ancestors as "bipedal chimpanzees." This may be a little unfair, but it's certainly true that, some details of the teeth and skull apart, there isn't a great deal more than postural erectness to link these early hominids to ourselves.

Given the age of the new form, even tenuous indications of uprightness (yet to be confirmed or denied by study of the new *Ardipithecus* skeleton) were bound to excite speculation on the nature of the environment within which bipedality evolved. The adoption of bipedalism followed a worldwide episode of drying and increased seasonality of rainfall that was initiated about 10 myr ago. This was the first of a series of climatic deteriorations that climaxed in the violent oscillations of what is known in more northerly regions as the Ice Ages. From this point on, we find evidence for the intermittent expansion of more open conditions at the expense of forests, with grassy woodlands becoming an increasingly important feature of the landscape (although the familiar Serengeti-type open grasslands familiar to tourists today were apparently not firmly established until quite late, perhaps even as recently as only 1 myr ago). At about 7 to 5 myr ago (the record is poor) we find an increase in the fossil record of medium- and large-bodied browsing and grazing animals that presaged the familiar fauna of today's eastern and southern African woodlands and savannas. Not everyone was convinced, but the obvious implication of these environmental changes was that humans had evolved from a population of hominoids that had been forced out of the closed forests by the stepwise wasting away of this ancestral forest environment. More about this later; meanwhile, it was noteworthy that the environment in which the known *Ardipithecus* individuals died had been quite densely wooded. Of course, most species tend to straddle a variety of habitats, and the one in which a given fossil is found may well not be typical for its species; but the observation seemed particularly significant in view of newly published reports of closed habitats for certain other early hominids.

Solving this problem for *Ardipithecus* took on a rather less urgent aspect when, less than a year after this new form had made its debut, Kenyan investigators announced the discovery of another early hominid, almost as old at 4.2 to 3.9 myr. This new species, *Australopithecus anamensis*, was assigned to a genus of early hominids that had long been known; and despite being a good half million years older than its successor species, *Australopithecus afarensis*, it looked comfortingly similar to the latter in comparable features. The small collection of *A. anamensis* fossils from the northern Kenyan sites of Kanapoi and Allia Bay contains decent upper and lower jaw specimens (one of each) with wide, thick-enameled chewing teeth, plus some isolated teeth and fragments of skull and limb bones. Among the latter is a partial tibia (shinbone) that shows pretty plausibly, especially in the bit that forms part of the knee joint, that its owner *had* walked upright. Fairly good evidence of bipedalism, then, back to about 4 myr ago. The ancient environments at Kanapoi and Allia Bay seem to have contrasted quite strongly with those at Aramis; at the Kenyan sites, riverine forest graded rapidly into woodland and quite open conditions. This bespeaks a relatively dry climate, and an environment much closer than Aramis to the one in which most paleoanthropologists had expected bipedalism to have evolved.

It is too early at this writing to predict the reaction of the paleoanthropological community, but clearly the discovery of *A. anamensis* has simplified the apparent anomaly in the human evolutionary record raised by *Ardipithecus.* These two forms are too close in time, and too different in morphology, for it to be argued at all convincingly that the vastly different *A. anamensis* is descended from *Ardipithecus.* The reverse, of course, is yet more implausible. In which case, we can for the moment charitably view the rather odd *Ardipithecus* as, at best, an early offshoot of the human lineage that lacks any known descendants. *Australopithecus anamensis,* on the other hand, lies much closer to what we see right now as the main line of hominid descent. Better knowledge of *Ardipithecus* may demand revision of this scheme; but whatever the case, the *A. anamensis* fossils do appear to project essentially *A. afarensis*–like morphology back into the past by an extra half million years.

Which brings us back to *Australopithecus afarensis,* the earliest hominid species for which we have a substantial body of evidence. In the mid- to late 1970s, hominid fossils were discovered both at Hadar, in Ethiopia, and at Laetoli, in Tanzania, that were—a few rather indeterminate fragments apart—older than any other hominids then known. Laetoli is most famous for the astonishing hominid trackways dated to about 3.6 million years ago—about which more in a moment—but it also produced a couple of nice partial jaws (one a juvenile) and various other bits and pieces. Hadar in its turn has by far the lion's share of anatomical fossils. The best known of these is "Lucy," some 40 percent of the skeleton of a small, probably female, individual. She is, however, run close in significance by the "First Family," a dazzling array of several hundred fragmentary fossils that represent the remains of at least thirteen individuals, all found jumbled up close together in the sediments that enclosed them. It seems quite possible that all of these unfortunate individuals perished together in a single event, maybe a flash flood, in which case we have here the remains of a single social group (which would confirm that all the fossils belonged to the same species, despite the great size differences among them). There are many other hominid fossils from Hadar, too; all fell within the short span from about 3.4 to 3.2 myr ago, until the recent discovery of a large and fairly complete cranium (as yet undescribed) about 3 myr old. Some bits from the Middle Awash, not far away, take the time range of these Ethiopian hominids back to about 3.8 myr.

A notable characteristic of the Hadar hominids was that while all displayed a relatively uniform bony morphology, there was an enormous size range among them. Early assessments tended to the view that two hominid species were represented there; but most paleoanthropologists are now content to accept at least provisionally that, along with the Laetoli specimens, all represent the single species *Australopithecus afarensis*. This species is strongly sexually dimorphic; in other words, as in gorillas, for example, the males were much larger than the females. It's been estimated that males weighed something like one hundred pounds, while females came in at only about sixty-five pounds. If the "Lucy" skeleton was typical, females stood only about 3½ feet tall; males were obviously taller, though it's uncertain by exactly how much (maybe a foot gets us into the ballpark). A male skull reconstruction, cobbled together from bits and pieces found at different Hadar localities and confirmed in its essentials by the new skull find, boasted a braincase that fell within the modern-ape size range and was coupled with a large and quite strongly projecting face. This skull thus shows rather apelike overall proportions. But the chewing teeth of *A. afarensis*, though big and mounted in jaws of decidedly primitive proportions, have some features in common with later hominids (including thick enamel). Also, the canine teeth are smaller than those of apes, though they are stout-rooted, and larger and a bit more projecting than ours.

Where *A. afarensis* really scores, though, is in the body skeleton. The very first discovery at Hadar was of a knee joint that showed typical characteristics of an upright biped, and the "Lucy" discovery amply confirmed that here was an erect walker. Early claims for the excellence of "Lucy's" adaptation to upright striding after our own fashion have since been disputed, but the basic adaptation was certainly there, both in the pelvis and the lower limb bones and in aspects of the upper body that correlate with this posture (including, in other specimens, a somewhat forwardly shifted foramen magnum). It's natural enough that a very early biped would not have been a perfect reflection of ourselves, and it seems that "Lucy" and her kind had, for example, longish arms and shortish legs, rather short upper arm and leg bones, somewhat long and curving hands and feet that were built for strong grasping, and narrow shoulders: all holdovers from a tree-living past. All of these primitive features would have compromised the efficiency of bipedal locomotion, and the suggestion has been made, for instance, that *A. afarensis*, with its long feet, was a rather poor runner. Nonetheless, the conclusion that *A. afarensis* was bipedal on the ground is reinforced by the Laetoli footprints, already mentioned, which provide an impressive direct demonstration of this kind of locomotion.

At Laetoli's Site G, two parallel sets of prints, left 3.6 myr ago in wet volcanic ash, proceed in undoubtedly bipedal fashion over a distance of almost eighty feet. Inevitably, the exact character of these prints is argued over, though it seems a good bet that their makers did not walk exactly as we do. This conclusion is reinforced by some 3.5-myr-old foot bones recently described from South Africa's site of Sterkfontein that indicate the individual in question had possessed, somewhat like apes, a divergent great toe with grasping capabilities, rather than an in-line one like ours (though the rear of the foot was more humanlike). But there's no question that the Laetoli footprints (presumably made by *A. afarensis*, though, of course, we can never be sure) amply confirm that hominids were up and walking on their hind feet by this early stage, even though how efficiently they did so remains open for debate.

A mosaic of primitive and more advanced locomotor characters seems intuitively reasonable for a hominoid that had not too long before abandoned an exclusively forest-living existence. Life would have been dangerous away from the forest for small-bodied and relatively slow-moving creatures such as these, and it would be surprising if they had not regularly sought the shelter of their former home. "Lucy" and her like were equipped with a basic adaptation that has subsequently proven, albeit in refined form, highly efficacious out on the savanna. Yet they had retained body features that would have made it relatively easy to move around in the trees in search of food and protected sleeping sites—though they would obviously have been far less accomplished arborealists than the apes. The environment at Hadar was one of riverine forests grading into evergreen bushland and savanna: not too far, perhaps, from what you might expect for a hominid of this kind. At Laetoli, though, the immediate habitat in which the footprints and fossils had been preserved was one of relatively open grassland, with few trees. But even this is evidence, if you wish to read it this way, that early bipeds used their new posture in order to move across hostile grasslands between shrinking areas of more congenial habitat: the tracks were headed toward the Olduvai Basin, in which forest surrounded an ancient lake, and the fauna associated with the fossil hominids indicates that plentiful trees were not too far away.

Still, the question remains: Why bipedalism? After all, plenty of four-legged animals do perfectly well out on the savanna. Suggestions have abounded. By standing upright, you look larger (and these early hominids were very small and vulnerable to large predators). What's more, carnivores are often "triggered" by a horizontal silhouette in a way in which they aren't by a vertical one. By moving upright, you also extend your horizon and you can constantly see over long grass to scan for predators. If you have your hands free, you can carry things, although what is uncertain if it wasn't food—most forms of which are not portable in any quantity without a container. Calculations of the relative efficiency of moving around in open areas bipedally versus quadrupedally have come up with mixed results, although it seems that when on the ground, you may make energetic savings by taking the bipedal route if you are already a hominoid with a tendency toward holding your trunk erect. Interestingly, though, it has recently been noted that chimpanzees tend to adopt bipedal postures more frequently when they are foraging in the trees than when they are on the ground. This, taken in conjunction with the emerging realization that the habitats represented by many very early hominid sites were far from as open as had been expected, suggests that the initial adoption of bipedalism among hominids was not propelled by the exigencies of the open savanna. Vast swaths of southern Africa today are covered by grassy woodland and bushland, often with quite large and closely spaced trees. This is a stable environment, not one that is merely transitional between forest and open grassland; and it is a habitat in which early hominids would have been able to make the most of their varied locomotor abilities.

Nonetheless, the savanna eventually became a factor, and a current intriguing suggestion as to the advantage of bipedalism out in the open (or even in only partial shade) invokes not locomotion as such, but rather the control of body temperature, and particularly that of the brain. Brains are delicate instruments and are particularly sensitive to temperature. If your brain overheats for more than a moment, you are a goner; and moving from the shade of the forest into the broiling heat of the tropical savanna or even of woodland means that you somehow have to keep your brain cool. Most savanna-living mammals have specialized structures to cope with this problem; hominoids don't. The only way a hominoid can keep the brain from overheating is to keep the entire body cool; and by standing upright, you dramatically reduce the area of your body that is directly exposed to the sun's vertical tropical rays. What's more, at the same time, you increase the body area available to lose the heat absorbed from the sun and generated by metabolic processes; and once the bulk of the body is high off the ground, it can also be cooled by blowing breezes. This latter mechanism works particularly well if, as humans have done, you have lost your heat-trapping hairy coat (except on the exposed head) and have developed an extensive network of sweat glands that enhance heat loss through evaporation. What's more, although upright locomotion may not be particularly fast, since you have enhanced your ability to lose heat by adopting an erect posture, it has the advantage of sustainability: humans have incomparable stamina and can remain active in the heat of the sun for long periods.

However, given the numerous compromises with committed upright locomotion that retaining tree-living adaptations involves, it seems probable that what we see in *A. afarensis* is an exaptation to savanna life: something that arose in a forest-to-woodland context and that only much later proved advantageous (in somewhat modified form) on the savanna as that new environment expanded. As I suggested in the last chapter, it is probably fruitless to try to identify *the* advantage conferred by upright bipedalism; once you stand upright, all of the potential advantages of this strategy are yours. But what's also true is that not all of those advantages need have been exploited at once. And the *combination* of locomotory features seen in *A. afarensis* may well suggest that they were not. Only a more thorough knowledge of the environments exploited by the earliest bipeds will help us sort this one out.

Whatever the facts of the case, this combination seems to have produced a winner, even though adapting a quadrupedal skeleton to upright posture has a definite downside that includes slipped disks, dislocated hips, wrenched knees, fallen arches, and a whole catalog of associated woes. Bipedalism in its "primitive" form was established in our lineage by over four million years ago; and following this, we see virtually no change for over half of subsequent hominid history. This "have your cake and eat it" locomotor strategy of early hominids was a stable and thus almost certainly a highly advantageous one: one that allowed its possessors to exploit the food resources and the protection offered by the trees even as, simultaneously, it offered them—at least in potential—the ability to cope with the stresses of more open conditions and to exploit the opportunities offered by this expanding habitat.

## *Early Hominid Lifestyles*

So, *A. afarensis* was a small-brained, large-jawed hominoid that was able, to some degree at least, to fend for itself out in the open as well as in more wooded environments: a "bipedal ape" indeed. What else can we say about this equivocal creature? Less, alas, than we would like. Claims have been made, though, that the adoption of bipedalism entailed considerable social consequences. Human females have abandoned the ancient estrous cycle of the apes, whereby ovulation is signaled by the swelling of the perigenital area. In the process of standing upright, the female genital region becomes hidden even as the chest is exposed, and it is probably no coincidence that human females have acquired prominent breasts and reputedly continuous sexual receptivity, even as ovulation remains cyclical and externally unannounced. On the basis of this, it has been suggested that the society of *A. afarensis* (for social this creature must certainly have been) was based on monogamous bonding of males with females. By consistently exhibiting alluring sexual signs (the breasts) and hiding ovulation, females attracted the constant attention of males, who could help them with the costs of rearing offspring by carrying food to them using their newly freed hands. In return, the females offered the males the certainty that the offspring they supported were their own; and the males responded appropriately.

In substantiation of this scenario, it was noted that in *A. afarensis* the canine teeth of (larger) males and (smaller) females are of similar relative size: something that contrasts with the "promiscuous" great apes, among which males have much larger canines than females do. And similarity between the sexes in canine size is typical of higher primates in which pair-bonding occurs. So far so good; but here we run into the basic problem of using analogies of this kind: which analogy do you choose? For if *A. afarensis* was indeed a single species, it was one in which the males were very much bigger than the females; and sexual dimorphism of this degree is typical of those apes among which males aggressively compete for females. What's more, a comparison of social traits in the various species of living hominoids has suggested that the common ancestor of hominids and the extant apes would have enjoyed a closed social system in which males were dominant and mutually aggressive, females transferred out, and mating was promiscuous. This notion is also supported, if weakly, by ecological considerations. Sadly, though, the truth is that as long as we have to rely on second-order inferences such as these, we will never certainly know in exactly what kind of social milieu our earliest ancestors evolved.

We may, however, be able to hazard some suggestions about the cognitive attributes of *A. afarensis.* If the ancestral retentions of this species mean as much as its anatomical and locomotory novelties (and the fact that its basic "have your cake and eat it" structure persisted for so long implies that they do), and if—less certainly—the preferred habitat of these creatures was broadly comparable to that of the woodland chimpanzees of today, then we have no strong reason for inferring that much of a cognitive gap existed between the early hominids and the living apes. To claim otherwise with any confidence would require a more refined knowledge of the ecological niche(s) of *A. afarensis* and of how this species exploited it than present evidence provides. Such knowledge will, with luck, eventually become available, although exactly how is now unclear. But until then, it is probably most prudent to view these early hominids much more as "bipedal apes," in cognitive as well as anatomical qualities, than as dim reflections of ourselves.

Still, it's hard to resist speculating a little. For example, the modification of the pelvis involved in bipedal anatomy presents human mothers with a unique difficulty: one which has implications for social cooperation. While the birth process in quadrupedal monkeys is less simple and easy than has generally been believed, the neonate twists during its passage through the birth canal to face the mother, who can thus assist in its final emergence. In humans, on the other hand, the baby has to twist to face away from the mother, who thus cannot provide such assistance for fear of breaking its back. Neither can the mother attend by herself, as monkeys can, to clearing mucus from the baby's nose and mouth to allow it to breathe or to unwinding the umbilical cord from around the baby's neck. All these attentions are frequently necessary, which is why midwifery is virtually universal in human societies. It has been suggested that the involvement of females other than the mother in the birth process goes right back to the origins of bipedalism; and if so, this implies a level of cooperation and coordination among early hominid females that goes far beyond that involved in the occasional infant care by "aunts" seen in other primates. Inferential again, and subject to better knowledge of the birth process among apes—but certainly suggestive.

How these early hominids made their living is also a matter of speculation, although even if their preferred environment was a woodland one rather than open savanna, it was a dangerous place to be. Grassy woodlands support numerous browsers and grazers, and along with them the predators—hyenas, leopards, lions—that prey on such animals. Given this omnipresent danger to the small, slow bipedal early hominids, there must have been considerable countervailing attractions in the form of new resources to be exploited. Among these must have been the tubers and rhizomes of ground plants as well as fruits and young leaves in the trees; and it is reasonable to assume that the early hominids sought the former out, perhaps using sticks to penetrate the hard earth. Chimpanzees, as we've seen, regularly hunt small mammals, and the early hominids presumably did so, too. None of this, of course, takes the hominids beyond the chimpanzee cognitive league; but it may be significant that the woodlands extended opportunities for scavenging the carcasses of dead animals (something which, as we've seen, chimpanzees don't usually do). Carcasses would have provided an unprecedentedly rich and compact source of highly nutritious proteins and fats, although the hominids would have faced stiff competition for them. It's even been suggested that the early bipeds capitalized on their retained arboreal abilities to steal the carcasses that leopards typically leave stashed in trees while they patrol their territories.

The addition of animal fats and proteins to the diet would have added an entirely new dimension to the activity of foraging. In the forest, fruits and other foods change seasonally but nonetheless are quite predictably distributed within the relatively small ranges of primate groups. Carcasses out in woodlands and savanna fringes, in contrast, are relatively randomly scattered around over a huge area (although carnivore kills are often most abundant near water sources). To locate the remains of dead animals in such environments, it is helpful to be able to read indirect signs, such as vultures clustered in a tree or wheeling in the air above. If one wishes to go beyond the direct record (which neither confirms nor excludes the possibility), then it is possible to guess that the early hominids on the savanna were already reading such evidence; and, if they did, they must have had commensurately improved cognitive capacities.

Still, stone toolmaking and language remained far in the future, although given what we know about chimpanzee tool use, it seems likely, as I've said, that the earliest hominids used digging sticks and other soft implements. They may have used rocks to smash nuts and other hard food items and may even have hurled stones (something too often underappreciated as a feat of perceptual and neuromuscular coordination) to discourage competition for carcasses. But even in cases where such activities might have left material evidence behind, it was not of the kind that preserves over the millennia. For the moment, then, there's little that we can say for sure about the lives of our earliest forerunners. But, whatever the exact lifestyle of the first hominids, it was clearly a successful one. For if the lack of significant functional change in the early human fossil record means anything in this context, this lifestyle endured for a very long time. Involving as it did a varied existence led between the forest fringes, woodlands, and open grasslands, it bespeaks considerable behavioral flexibility on the part of our remotest forebears: something that has apparently characterized hominids ever since.

## *South African Early Hominids*

In the period between about 3 and 2 myr ago, the focus of paleoanthropological investigation shifts to South Africa, whence in the 1920s the first species of *Australopithecus*, *A. africanus*, was reported. A little more lightly built than *A. afarensis*, this species was functionally quite similar: a small-brained, projecting-faced archaic biped unaccompanied by any evidence for tool use. An as yet undescribed partial skeleton, about 2.5 myr old, is said to be at least as primitive in its body proportions as "Lucy," and the 3.5-myr-old partial foot I've already mentioned shows a strongly divergent great toe, evidently capable of powerful grasping. Anatomies such as these fit well with the recently recognized presence of forest lianas and other evidence of closed conditions at *A. africanus* sites.

Early on, the broken nature of the mammal (including *Australopithecus*) bones at some of these sites was interpreted as evidence for hunting and even cannibalism on the part of our early relatives, inaugurating the "blood-spattered, slaughter-gutted archives of human history," as the anatomist Raymond Dart once wrote. The reality is more prosaic than this "killer ape" scenario, however. Subsequent work has shown that the bones recovered at the South African sites (which consist mostly of underground cavities into which bones fell from the surface) were actually scavenged by porcupines or were otherwise collected by natural agencies. Indeed, the very first *Australopithecus* discovery (a juvenile) appears to have been the victim of a giant eagle; and at another South African site, there is convincing evidence that the hominid remains represent the leftovers of leopards' meals. The hunters have become the hunted and, in doing so, have shed a lot of their assumed cognitive apparatus.

*Australopithecus africanus* is far from the only early hominid known from South Africa. Another branch of the human family is represented there, too, by the genus *Paranthropus:* a more massively built hominid, with huge chewing teeth that form a powerful grinding apparatus. Only a few bones of the body skeleton are known (enough, though, to show that *Paranthropus* was bipedal), so it's unclear whether these "robust" hominids were physically much larger than the "gracile" *A. africanus.* It's pretty obvious, however, that they must have had a more specialized vegetarian diet than the latter, which retained a more generalized dental apparatus. The brains of *Paranthropus* were on average a tiny bit bigger than those of *A. africanus*, but this difference may possibly be accounted for by larger body size. At one relatively young site, some ungulate horn cores and bone fragments have been found that show a curious polish, bearing witness to their use as digging implements; but this is the only clear-cut evidence that any truly archaic hominids ever used tools.

The picture of early hominid variety in South Africa will certainly become more complex as the human fossil assemblages from the half-dozen early sites in the region come under increasingly intense scrutiny. It is looking ever more probable that different "robust" species are represented at the two main *Paranthropus* sites, and an early form of *Homo* at one of them; and there is enough morphological variation at the classic *A. africanus* sites to suggest that at least one other hominid species is represented in addition to that one. Historically, then, South Africa will always retain its singular importance as the area from which truly early hominids were first reported; but currently its foremost significance is in providing us with some of the best evidence for hominid species diversity in the period between 3 and 1.5 myr ago. Given that for many years it was fashionable to view human evolution as a linear process, that undeviating slog from primitiveness to perfection, this is salutary indeed. The picture emerging from our new appreciation of human evolution in those early years is one of variety: of numerous evolutionary experiments, most of which ultimately failed. The roles of speciation and extinction in human evolution have come to the fore, bringing the documented history of Hominidae into line with that of other mammal groups and erasing, hopefully forever, the notion that there was anything particularly special or unusual about how we ourselves came into being.

## *The First Toolmakers*

At about 2.5 million years ago, there occurred a dramatic shift in the nature of the African fauna. Another climatic deterioration, marked by an increase in polar glaciation and global aridity, spurred the ongoing conversion of forest to woodland and woodland to savanna. The fossil record reflects this change; grassland antelope species, for example, appeared in great abundance even as woodland species diminished. At this time, too, stone tools first show up, inaugurating our archaeological record, and we also find the first fossil intimations of our own genus *Homo*. It's widely believed that these developments were related, our emergence and the invention of technology both representing responses by the ancestral human population to the climatic and ecological vicissitudes of the time.

It is difficult to summarize the events in our ancestral lineage in the period from about 2.5 to 1.5 million years ago, partly because the definition of *Homo* and its contained species is so poor and partly because paleoanthropologists have yet to make up their minds about what the mainly rather fragmentary fossil record over this period is telling us about the variety of hominid species that then existed. The first truly ancient species of *Homo* to be described (in 1964) is *Homo habilis*, named on the basis of some 1.8-myr-old fragments found at the bottom of Tanzania's Olduvai Gorge. These were thought to represent the maker of the rudimentary stone tools that had long been recognized in the lower levels of the gorge; and it was largely under the then-influential concept of "Man the Toolmaker" that the bones were assigned to a species of our own genus, for anatomically they did not in the main differ hugely from *A. africanus* (although the original *habilis* specimen had a rather larger brain). Finds made in the 1970s, principally in Kenya's East Turkana region, complicated the picture considerably. Here, stone-tool–bearing sediments 1.9 to 1.8 myr old produced a range of hominid skulls, some of them small brained and reminiscent in certain respects of *A. africanus* and others (notably the famous ER-1470 cranium) with brains half as large again. A few isolated limb bones also suggested that two types of hominid were here: one with more archaic, the other with more modern, body proportions.

Work at Olduvai in the 1980s also produced a fragmentary but spectacularly archaic skeleton of about the same age as the original *habilis* find. Inevitably, there is considerable ongoing argument over how many species are represented in the Tanzanian and Kenyan fossil assemblages (which also contain "robust" hominids), over which specimens should be allocated to which species, and what the species concerned should be called. This argument is far from resolution; but it is at least possible to speculate that in both regions, two species existed in this time range: an archaic form broadly comparable to *A. africanus*, and *Homo habilis* (including ER-1470), with a larger brain and somewhat more modern body proportions. If so, it seems reasonable to conclude, if impossible to prove, that *Homo habilis* was the toolmaker in both regions. A lower jaw from Chiwondo, in Malawi, may extend the time range of a broadly defined *Homo habilis* back to about 2.4 myr—which is, interestingly, about the time at which the earliest known stone tools appeared—at other sites in Kenya and Ethiopia that unfortunately lack fossils.

It is unfortunate that we know so little at present about the physical attributes of the first known stone toolmakers, but this situation should improve in time. Meanwhile, how about the tools themselves? I emphasize "known" because what I have been talking about is actually the first *knowledgeably made* stone tools; and archaeologists are unsure about what might have preceded the making of stone tools of this kind. Their science went through a difficult period in its early years, when naturally broken stones were thought to have been tools; and having learned some hard lessons from this, few archaeologists today, for example, would willingly recognize any of Kanzi's "artifacts" as tools from their form alone, without more information. In any event, the first *recognizable* class of stone tools is that associated with *Homo habilis*, the first examples showing up at about 2.5 million years ago. And despite their simplicity, they tell us quite a bit about the capacities and activities of their makers.

Not that they are too impressive to look at. They consist for the most part of sharp flakes banged from smallish cobbles using a stone "hammer." It was once thought that the "shaped" cobbles themselves were the primary tools, and they were classified into a number of distinct forms; but experiments have demonstrated pretty convincingly that these "cores" were principally, at least, by-products of the flake-making process. Rudimentary though a stone chip a couple of inches long may seem, though, it can be a surprisingly efficient cutting tool, especially when made from the right kind of rock. Experimental archaeologists have shown that an entire elephant can be butchered quite rapidly using such utensils. What's more, making usable flakes takes not only advanced manual skills, but a comprehension of the intrinsic properties of the material being worked and of the principles on which it can be induced to fracture appropriately. It takes a good bit of cognitive sophistication to work out how to bang one stone against another at precisely the right angle to produce a usable flake; and as we've seen, this task appears to be well beyond the abilities of even the smartest chimpanzee. No matter how modestly sized the brains of the first stone tool-makers may have been, their owners had moved cognitively way out of the ape league, though just how far remains a matter of inference.

Perhaps even more significantly, the record contains evidence that stone toolmaking was not simply an opportunistic process carried out as necessity or convenience dictated. The makers of these early "Oldowan" tools (named for Olduvai Gorge, where they were first identified) not only knew which rocks were most suitable for toolmaking, but often carried such rocks around in anticipation of needing them. We know this because sites at which early stone tools have been found are frequently quite far away from the nearest natural sources of the rock used for making the tools and because researchers have repeatedly been able to piece together complete cobbles from fragments—both tools and waste pieces—collected at butchery sites. The stone was worked where it was used, which was not where it came from; and it seems improbable that the hominids involved would routinely have found (or, much less likely, killed) a large mammal and only then have traveled several kilometers in search of appropriate materials with which to butcher it. Positive evidence for the butchery of carcasses using the stone flakes found along with them is provided by characteristic cut marks left on the bones by stone tools during dismemberment and the slicing away of hunks of meat.

Butchering a carcass out on the savanna was clearly a hazardous proposition for small-statured early hominids since there was plenty of competition for such resources. Quite apart from such top predators as lions and leopards, there were dangerous hunter/scavengers and some of the larger ungulates to be contended with (hippos and lone male buffalo, especially, are not to be trifled with—even by modern humans). Such beasts as hyenas and hunting dogs would have shown considerable interest in any even marginally meaty mammal carcass, and early hominids would have had to discourage interest of this kind, at least until their tools had allowed them to detach parts of the carcass for transportation to a safer place for consumption. Stone flakes would have afforded little defense against a ravening hyena, and something more would have been required. What might it have been? Chimpanzees have been seen to flail dead, stuffed leopards and other intruders with sticks, but these would have been a resource in short supply on the savanna. Stones for throwing abound there, on the other hand, and it's also known that chimpanzees throw objects at intruders, though they are not noted for distance or accuracy. We'll never know for sure, but it's quite plausible that early hominids defended themselves on the savannas by hurling objects at unwanted visitors. An early hominid whose eye-hand coordination allowed him to throw hard and accurately against a large-fanged competitor would certainly have been at an advantage relative to one who couldn't. And, interestingly but possibly not coincidentally, it turns out that the most painless way of making a simple stone tool by banging one rock against another is to avoid holding either too tightly and instead to "throw" the loosely held hammer stone against the core being flaked.

Given the rather undignified reputation that attaches to human scavenging today, early archaeologists tended to favor the notion that our early stone-tool–using ancestors had been hunters of medium-bodied and larger mammals, compensating for their diminutiveness and frailty by guile and intelligence. This outlook may have gratified the modern human ego of its authors, but it has the profound disadvantage of interpreting the past in terms of present experience: a deeply flawed approach. Nowadays the notion of active early human hunting of sizable mammals has largely been discarded, and attention has been directed toward the kinds of scavenging in which those ancestors may have indulged. Hominid scavenging is actually quite a demanding activity, although it was certainly facilitated by the fact that predator-devoured carcasses tend to lie undisturbed longest along the edges of riverine forest, precisely the kind of habitat that was probably favored by those remote ancestors.

Our best evidence for the activities of the early human toolmakers comes not from the individual artifacts themselves, but rather from the nature of the sites at which they are found. Oldowan localities consist of spots on the landscape at which early humans left evidence of their activities; and such evidence usually comes in the form of accumulations of stone tools and broken animal bones. The problem is that animal bones are regularly broken by agents—lions, hyenas, and leopards, for example—that have nothing to do with human activities. So when you find ancient bones that were broken before they were fossilized, it's important to scrutinize them very carefully to determine the kinds of markings that were left on them; hyena teeth, for example, scar the bones in a different way from human tools. But even where you find cut marks on bones that were undoubtedly made by tool-wielding humans, there's no way to determine whether the animal in question was actually dispatched by those humans. It might equally well have been killed by a large predator and later scavenged by hominids: physically delicate creatures who may have often had to let other scavengers go first.

This doesn't mean that it's impossible to say anything about the sequence of events at a kill site or some other place where hominids may have had a hand in bone accumulation. In some cases it's possible to make probability statements about what happened. Let's take one example. Field experiments have shown that large carnivores such as lions devour the most readily available portions of a carcass. When they've finished, only about 15 percent of the bones show tooth marks, the entrails and fleshy areas having captured most of their attention. When hyenas have followed (or achieved the kill in the first place) with a much more thorough job of consumption, some 80 percent of the bones will display dental scarring, although usually the midshafts of the limb bones will be unmarked. So if at an archaeological site the animal bones show only about 15 percent tooth marks and also display distinctive cut marks (V-shaped in section) made by stone tools, the probability is that hominids had followed the lions but had done their work before hyenas had arrived. However, even when hyenas have been at a carcass—and vultures have finished their external cleaning job—there is still a resource of interest to hominids. The neglected long-bone shafts are a rich source of highly nutritious marrow for any creature able to break open the bone. This can be achieved by pounding the shaft with a rock. An attack of this kind typically produces an unusual kind of torsional fracture along the bone, revealing the marrow within: and such fractures are not uncommon among animal bones found at Oldowan sites.

What's more, the proportion of limb bones at some Oldowan localities is unexpectedly high, suggesting that these elements had repeatedly been transported to those sites from the places at which the complete carcasses had originally lain. This implies that such sites were originally in sheltered spots where the hominids habitually consumed marrow, away from the dangers that attended in more open areas. None of this means, of course, that early tool users were necessarily specialized marrow eaters. This is simply an activity for which the evidence is of a kind that tends to be preserved in the geological record. Nonetheless, while the probability still must be that the diet of the first toolmakers was quite varied and mostly vegetarian, the exploitation of marrow was a totally unprecedented behavior for hominids, and one that opened the way for other innovations to come.

Early studies of Oldowan localities resulted in the suggestion that they were "living sites," a notion later expanded to that of "home bases," centers to which the hominids habitually returned from foraging ventures. From this it was but a short leap to view these places as the focal points of a complex lifestyle that involved the sharing at such central locations of food acquired in the region around and carried back in hands unoccupied by locomotion. In turn, this implied complex communication among group members (which might even have amounted to language), reciprocity in social relationships, and division of labor between the sexes (for females, burdened with children, would not have hunted but instead have gathered plant foods around the camp). The list of such putative behaviors went on; but it was soon realized that this perspective—which was based on the assumption that if they didn't behave like apes, they must in some way have behaved like modern humans—leaned excessively toward viewing the early hominids in our own image. Archaeologists today are considerably more circumspect in their interpretations.

Nonetheless, for the first time it's abundantly clear that early hominids were significantly more capable than chimpanzees on a whole variety of scores. Beyond the aspects of toolmaking and tool using I've already mentioned, we can infer from the exact configuration of the flakes produced in the toolmaking process that, like us, the toolmakers were mainly right-handed, which implies some brain reorganization relative to apes. This ties in nicely with the observation that the endocast of ER-1470 is the earliest known to show any notable asymmetries between the two sides of the brain; and, as we've seen, such asymmetry impacts on a lot more than just handedness. Exactly what this implies about the cognitive abilities of ER-1470 is, though, impossible to specify. We don't know, for instance, how well such hominids communicated with each other, how large or complex their social groups were, how much insight they had into each other's states of mind, or how capable individuals were of interpreting their own motives.

Still, it remains probable that finding food out on the savanna would have demanded of these hominids a subtlety of observation and mental association surpassing that of full-time leaf or fruit eaters in forests. But even though these early toolmakers were undoubtedly able to exploit mammal carcasses in a much more effective way than their precursors could ever have done, it remains true that we have no way of knowing how much life had changed in other respects. There's no question that with the invention of stone tools we are witnessing a major hominid lifestyle change, as well as a cognitive innovation of enormous consequence. Nonetheless, it would be profoundly misleading to think of these toolmakers simply as "primitive" versions of ourselves. And I doubt very much whether, if through some miracle we could meet them in the flesh, we would intuitively wish to describe much about them as functionally "human."

CHAPTER 5

# *Becoming Human*

We saw in the last chapter that our earliest African ancestors were in some sense "compromise" creatures. They had left the depths of the forest, but they had not committed themselves to the open savanna. They were less expert arborealists than the living apes, though handier in the trees than we are; and while more at home on the ground than the apes, they were less efficient bipeds than later humans. It would be a mistake, though, to assume that they were somehow stranded between two alternative lifestyles, for their adaptation proved a highly successful one: one that served them well for two million years. But eventually something happened that precipitated one group of them away from the woodland habitat that had been their home, and out, definitively, onto the open savanna. That something may well have been an intensification of the trend toward aridity and seasonality of rainfall that had been afflicting Africa for some time, with consequent accelerated conversion of grassy woodland to open grassland. It is, of course, frustrating that we are at present so far from understanding exactly what our ancestors of that transitional period were like physically, though more careful analysis of the currently available fossils, as well as future discoveries, will with luck help us comprehend this better. But even now we do not have to wait long for dramatic evidence of creatures in whom we can, physically at least, recognize at least some of what we see in ourselves today.

## *Toward Modern Humans*

During the middle 1970s the ever-productive East Turkana region of Kenya turned up evidence of a new kind of hominid in the 1.9-to-1.4-myr range—coinciding at the high end with *Homo habilis* in eastern Africa. Here was something distinctively different from anything previously known. Increasingly referred to as *Homo ergaster* (sometimes as "African *Homo erectus*"), this new species sported a cranium that much more closely resembled that of later humans, boasting a braincase of about 850 ml in volume (above half the modern average), housed in a relatively high and rounded vault. Its face was comparatively lightly built, though still somewhat projecting, and its chewing teeth were significantly smaller than those of earlier *Homo*. These cranial fossils were spectacularly complemented in 1985 by the discovery, in West Turkana, of a remarkably complete skeleton (the "Turkana boy") of a *Homo ergaster* adolescent who died 1.6 myr ago.

Analysis of this skeleton produced numerous surprises. The young individual (who probably died at about nine years of age, although his permanent teeth had erupted to a degree comparable to a modern twelve-year-old) had stood about five feet four inches tall. Had he lived to maturity, however, it's estimated that he would have reached an astonishing six feet. What's more, his skeleton was radically different from that of earlier hominids, possessing most of the hallmarks of modern human bodily form. Indeed, his overall proportions (though not relationships) are remarkably close to those of the slender-limbed people native to the Turkana region today. The modern Turkana, with tall, slight frames that maximize the heat-shedding surface areas of their bodies while minimizing their heat-producing volumes, are ideally suited to the arid tropical environment in which they live; and it appears that so, too, was the Turkana boy. You can even glimpse this, perhaps, in the shape of his nose, which possesses a projecting nasal bridge that may bear witness to an adaptation of the nasal passages to preserve moisture and to humidify dry incoming air. What's more, in *Homo ergaster* we find a continuation of the trend toward cerebral asymmetry that we first encountered in ER-1470.

Finally, then, here was an ancient hominid clearly suited physically to life out in the open savannas, emancipated from the forest-edge environment in which his ancestors had flourished. Of course, Turkana boy did retain certain features of those ancestors, such as an upwardly tapering rib cage and aspects of the hip joint; but there's no doubt that his species had achieved an effectively modern skeletal anatomy. Nobody is going to claim that Turkana boy and his relatives depended on the forest in anything like the way their predecessors had.

Culturally, though, little appears to have changed by the time Turkana boy came on the scene—although studies of tooth wear in *Homo ergaster* suggest a greater amount of meat eating compared to *Homo habilis*, which shows the more vegetarian pattern of wear typical of *Australopithecus*. The stone tool kit and animal remains found at Turkana in the time of the earliest *Homo ergaster* do not, for example, differ significantly from those associated with *Homo habilis*. At first, this appears a little counterintuitive; but a moment's thought will contradict intuition. Why should a new species—which inevitably emerged out of an old one—necessarily bring with it a new technology? Whether physical or cultural, innovations can appear in only one place, and that's *within* a species. Any individual who bears a new genetic trait or invents a new technology (which is a very different thing) cannot, after all, differ too much from his or her parents. Of course, by the same token, it's hard to see how the huge physical gulf between *Homo ergaster* and *Homo habilis* could have emerged in a single event, and there's presumably much that we're missing here in the fossil record (although the discovery of a class of genes in which relatively small changes can have large developmental effects may well be highly relevant here). But it's pretty clear that the period following about 2 myr ago was an eventful time, indeed, in human evolution, perhaps precipitated once more by climatic vagaries. It's possible also that the invention of stone tools set in train a series of events that helped transform our lineage radically but in ways that we cannot yet discern, and we can only hope that future discoveries will help us learn what those events were.

A couple of hundred thousand years after fossils of *Homo ergaster* first show up, however, we do see a remarkable cultural innovation in the archaeological record. Up to that time (about 1.5 myr ago), stone tools had been of the simple kind that had been made for the previous million years or so, in which the main aim had probably been to achieve a particular attribute (a sharp cutting edge) rather than a specific shape. Suddenly, however, a new kind of tool was on the scene: the Acheulean hand ax and associated tool types, which were obviously made to a standardized pattern that existed in the toolmaker's mind before the toolmaking process began. Hand axes are large, flattish, teardrop-shaped implements that were carefully fashioned on both sides to achieve a symmetrical shape; and because of the multifarious uses to which they were evidently put, they have been described as "the Swiss Army Knife of the Paleolithic." The Paleolithic, by the way, is the period more colloquially known as the Old Stone Age: the long era (that ended only about ten thousand years ago) within which stone tools were not only made but finished by fracture, rather than by polishing. With the Oldowan and the Acheulean, we are in the Early Paleolithic period; middle and late periods followed.

Microscopic examination of the edge-wear on Acheulean hand axes has shown that these tools served such functions as cutting, scraping, and hacking, making them true general-purpose implements. Nonetheless, the individual functions they served duplicated those of their predecessors. What's more, those predecessors didn't disappear as the new artifacts came in; Oldowan-type tools continued to be made for a very long time alongside the carefully crafted Acheulean implements. And, although evidence is quite sparse, there's nothing much to indicate that the lifeways of the new hominids had been altered substantially by the introduction of the new technology. Of course, the radically new concept (a mental template in the mind of the maker) that lay behind the Acheulean hints by itself at a cognitive advance on the part of the toolmakers, as does the fact that in later times such tools were often made in specialized "workshops," at which huge quantities of hand axes have been found. But at present there's precious little archaeological evidence to show that ancient hominid life had changed significantly as a whole—although on anatomical grounds, there's no doubt that *Homo ergaster* would have been a great deal more comfortable out on the grasslands than *Homo habilis* ever was.

Here we see an excellent example of a pattern that has consistently characterized the evolution of our lineage, in which anatomical innovation has proceeded independently of technological change: new species and new technologies are not directly linked. We'll run into more examples of this as our account proceeds. It is, however, during the tenure of *Homo ergaster* that we find the first intimations in the archaeological record of the use of fire—undoubtedly one of the most significant cultural innovations in hominid history. The earliest of these occurs in the form of burned cobbles and bones at a South African site thought to be about 1.5 myr old. At Chesowanja, in Kenya, burned clay balls indicate the occurrence of fire there in association with stone tools at about the same time, although whether the fire was natural or artificial is still debated. Clearly, the mastery of fire was a major event in human prehistory, carrying with it symbolic overtones as well as practical consequences in terms of protection against predators, cooking of food, and so forth. Certainly, it cannot fail to evoke echoes of humanity. The trouble is that these isolated cases fall a million years or thereabouts before we have any concrete evidence of the control of fire in hearths, so it is difficult to know exactly what to make of them. While stone tools, once invented, rapidly became a regular part of the hominid repertoire, fire use seems to have been rather intermittent until quite a late stage. We'll see this pattern repeated later in the hominid record, in other contexts.

## *Out of Africa*

At one time it was thought that it was only about a million years ago that hominids first penetrated beyond the continent of their birth to enter cooler climes to the north and east, propelled by who knows what curiosity or adaptability and cultural armamentarium. New dates now suggest, however, that humans had reached eastern Asia not long after we have the first evidence for *Homo ergaster* in Africa. Although the exact identity of these early Asians is not clear because the fossils involved are very fragmentary, it thus appears that the typically human wanderlust dates from the very origin of hominids of reasonably modern body form. The emancipation of hominids from the woodland habitat and their emergence as tall, striding savanna beings, it seems, also inaugurated their career as long-distance travelers. This doesn't mean, of course, that individual humans traveled vast distances or that expeditionary forces were sent out to explore new areas. But it does suggest that it was the colonization by humans of open habitats that cleared the way for the cycle of population expansion and geographical spread that culminated in the crowded world of today.

The climatic vagaries initiated by the ancient cooling and drying events I've already mentioned culminated in later oscillations of climate that most strongly affected the northern regions into which the original emigrants from Africa moved. Just as the first human diaspora began, world climate moved into a period of extreme instability that greatly affected subsequent events in hominid evolution. Before we continue the story of innovation in human physical and cultural evolution, let's pause briefly to look at this remarkable period of Earth's history, known colloquially as the Ice Ages.

## *The Ice Ages*

Early in the nineteenth century it was realized that the landscape of much of Europe has quite recently been altered by repeated glaciation during periods of intense cold. At such times, thick ice sheets formed and scoured the surface of the region. Thus was born the concept of the Pleistocene geological epoch, more commonly known as the Ice Ages. In the early years of our own century, geologists refined this notion to recognize four main glacial periods, separated by warmer interglacials, which were reflected in periodic expansions and contractions of the Alpine ice cap. This fourfold division of the Pleistocene, which is now reckoned to have covered the past 1.8 myr, persisted up into the 1960s, when new technologies began to reveal a much more complex picture.

Terrestrial studies of glaciation have always been complicated by the fact that each advance of an ice sheet tends to scrub away evidence of the glacial that preceded it, while the outwash from the melting of the ice cover with the coming of an interglacial has a similar erosional effect. On-land studies of the sequence of geological events during the Pleistocene thus often elicited more problems than they solved. During the 1950s, however, it was observed that the deposits that form at the bottom of the oceans preserve a much more continuous record of geological and climatic events than anything available on the continents.

The sediments of the seafloor are composed not only of mud washed into the sea from the land, but also of the remains of microorganisms that lived in the seas themselves. These tiny creatures, known as forams, absorb two different forms (isotopes) of oxygen from the waters around them and preserve a record of the ratio between them in their shells, which sink to the seafloor when they die. This isotopic ratio differs according to climate, since the lighter form evaporates preferentially from the seawater. In warm times the evaporated isotopes mostly reenter the sea through precipitation and runoff from the land; but during colder periods many are locked up in the polar and high-altitude ice caps. At such times the representation in the seas of the lighter, more readily evaporated isotope is reduced, making it rarer in comparison to its heavier counterpart. The accumulating deposits on the seafloor hence contain a record of climatic change, as the isotopic ratio in foram shells shifts in favor of the heavier form in colder times (when the forams also absorb it preferentially) and back again in warmer ones.

Examination of the isotopic record in cores taken from the seafloor indicated a much more complicated sequence of climatic events than had previously been thought. Over the past 1.8 myr, the isotopic signal has shifted between colder and warmer conditions approximately every 100 kyr. This has not, however, been a simple cycle. The major 100-kyr fluctuations have generally been marked by a relatively slow cooling followed by a rapid warming phase; but within each one there have been numerous smaller oscillations in world temperatures. What's more, the amplitude of climatic fluctuation has tended to increase, reaching a maximum over the last 120 kyr or so, a period corresponding roughly to the last major glaciation and the present warmer interlude. Whether this is a true pattern or one that merely emerges from our better understanding of more recent climatic change is yet to be certainly determined; but what we do know is that our Pleistocene ancestors lived in a period of extremely unsettled climates.

And climate is not the whole story. Certainly, as the world cooled and the polar ice caps expanded, vegetation zones moved southward (in the Northern Hemisphere) and to lower altitudes, while toward the poles ice cover inexorably increased, rendering previously congenial landscapes uninhabitable to humans. Animal herds upon which early humans depended moved with the vegetation and sometimes found their favored habitats disappearing or drastically reduced until warmer times returned. But in colder periods, ice in northern areas also covered routes otherwise available between areas favorable to humans, blocking access between them. And, most importantly, ice buildup radically changed world geography. As the major ice sheets swelled, they retained water that would otherwise have returned to the oceans; and sea levels consequently fell, by up to three hundred feet or even more. Dry land appeared along the continental margins, uniting islands with the mainland; and the reverse, of course, occurred when the climate warmed up.

Thus the initiation of the Pleistocene climatic cycle inaugurated a world of unstable geography and environments: one in which hominid populations were summarily separated from one another and were reunited equally arbitrarily (relative to any adaptations they may have possessed to prevailing circumstances). Such conditions are precisely those that, as we saw in chapter 3, are most conducive to evolutionary change. Varying ecologies force adaptive change in local populations, and the fragmentation and reuniting of populations accelerates the processes of speciation, competition, and extinction. It is out of this ecologically and geographically unstable world that our own ancestors ultimately arose. The widely cited notion of a monolithic "ancestral environment" that, through our genetic heritage, still conditions our behaviors today is simply unsustainable.

It's important to note, however, that although glaciation itself was confined to the higher latitudes during the Pleistocene, the effects of climatic fluctuation were global. Europe and western Asia were—and still are—peripheral areas of the entire Old World, throughout which human populations were spread during the Pleistocene. This is not to say that these areas were unimportant in human evolution: peripheral areas perform, as we've seen, significant roles in the evolutionary scenario. But Europe was not uniquely the stage on which the human drama was playing out; it is simply the area, as I've already stressed, from which we have the best evidence. Africa, too, was affected by climatic vagaries and their attendant evolutionary consequences; and although the evidence from this continent is lamentably sparse over the period of the later Ice Ages, it appears that major developments in human evolution occurred there. Nonetheless, it is from Europe and (in later periods) western Asia that we have the best record for this time; and in the next section we will explore this record.

## *Early Europeans*

The early non-African human fossil record is reasonably good in eastern Asia, where the species *Homo erectus*, a descendant of *Homo ergaster*, flourished for upward of a million years. In contrast, until recently human occupation of Europe appeared to have commenced fairly late. Now, though, early hominids have been reported from Spain as far back as about 800 kyr ago, which is about the date of the earliest reliably identifiable (if rather crude) tool assemblages from Europe. Their discoverers have assigned these early fossil fragments to the new species *Homo antecessor*, which they believe was ancestral to modern humans on the one hand, and to a lineage leading ultimately to the Neanderthals on the other. The *Homo antecessor* specimens are certainly more similar to *Homo sapiens* than is anything else in their time range; but it's early days yet, and it seems just as likely at this point that the remains are those of individuals who belonged to an early—and ultimately unsuccessful—group of invaders of the stressful habitats of Europe, as that they represent the direct forebears of later Europeans. And while the rest of this chapter will focus primarily upon Europe and western Asia, because that is where the record is best, it should be kept in mind that despite the general dearth of evidence, similar developments were presumably taking place around this time all over the Old World (the Americas were occupied by humans at an extremely late stage, arguably only by about 18 kyr ago). If there's one thing of which we can be pretty certain, it is that during the last million years there was a lot more going on in human evolution than we have yet been able to discern.

Conventional wisdom still rules, however, and conventional wisdom has had it that Europe was first successfully colonized by "archaic" members of our own species, *Homo sapiens*. This is now debatable from several points of view, principal among which is that (with the exception of *Homo antecessor*) the earliest well-documented human fossils from Europe belong—in my view, unquestionably—to an entirely distinct species (which may ultimately be found itself to require division), the proper name of which is probably *Homo heidelbergensis*. The Spanish researchers think that this species stands between *Homo antecessor* and the Neanderthals. Here, in the millennia following about half a million years ago, we are approaching a more familiar skull form, with brain volumes up in the 1,100 to 1,200 ml range, within striking distance of the modern average. The braincase still looks rather "poorly filled," though; and the face, overhung with largish brow ridges, is still hafted to the front of the braincase, rather than tucked beneath it, as it is in ourselves. As far as can be told from sparsely known bones of the body skeleton, these humans, while essentially similar to us in bodily structure, were very strongly built. Nor is this species a purely European phenomenon; the front of a massive skull of this species from Bodo, in Ethiopia, is maybe as much as 600 kyr old, and a cranium from Kabwe, in Zambia, possibly approaches that age. *Homo heidelbergensis* may also be known from as far afield as China, although the specimen involved, from Dali, is both poorly described and probably younger than those already mentioned.

Nonetheless, it is in Europe that *Homo heidelbergensis* is found in the best-documented archaeological contexts. A site of particular interest is the southern French locality of Terra Amata, which is about 400 kyr old and presumably the handiwork of *Homo heidelbergensis*, although no human fossils confirm this. Its excavators somewhat controversially interpret Terra Amata, now perched high on the corniche overlooking Nice, as a seasonally occupied beach camp to which ranging hunters regularly returned. Its most prominent feature is the earliest recorded shelters made by humans. The largest and best preserved of these consisted of an oval arrangement of saplings (not preserved directly, of course, but indicated by postholes) that was over twenty feet long. The saplings were planted in the ground and were, it's thought, bent inward to meet at the top. It's possible that this hut was weatherproofed with animal skins, although its chief excavator believes not. The periphery of the structure was reinforced by large stones, with a gap in the ring that indicates the entranceway. Inside, the floor was largely cleared of beach cobbles, although butchered animal-bone refuse was haphazardly piled around. Most significantly, the hut contained a shallow scooped-out hearth, containing burned stones, which is possibly the earliest strong evidence for the domestication of fire. This is a truly significant innovation, which makes it all the more surprising that evidence for hearths is so sparse over the following couple of hundred thousand years, although corroborating evidence comes from the German site of Bilzingsleben, not much younger than Terra Amata at about 350,000 years old.

The stone tools at Terra Amata (like those of Bilzingsleben) were pretty crude, although this may be due to the fact that they were made from silicified limestone cobbles: not particularly good raw materials for this purpose. It has, in fact, turned out that many variations among stone tools are due principally to the materials from which they were made, rather than to the level of technological sophistication possessed by the toolmakers. It may be significant, though, that the stone tools found in association with hominids at another southern French site, Arago, were hardly more magnificent, consisting for the most part of rather elementary "chopping tools" and large flakes. At Arago (a limestone cave entrance) a succession of "living floors" was found, with an abundance of stone tools and animal bones, and the site was evidently occupied by hominids, at least intermittently, over an extended period of time. Despite the animal bones found there, however, many archaeologists would be reluctant to characterize the Arago humans as skilled hunters. A lot of the bones might have been scavenged; and, more generally, there are many ways besides human activity in which fragmentary animal bones can find their way into archaeological deposits—they can be washed in by water, for example, or accumulated by hyenas.

Still, we have in sites of this kind the first plausible intimations of the "home base" of which recent studies have deprived earlier stone toolmakers. As we have seen, the existence of home bases implies a great deal about cooperation and social organization; and in such sites, for the first time, we begin to glimpse, if only dimly, some of the behavioral attributes that we most closely associate with ourselves. Yet, once more, specifics are elusive. Of course, the control of fire evident from the hearths at Terra Amata must certainly have made a difference in the inhabitants' experience of daily life. Fire provides light, warmth, and protection; it can be used to harden both stone and wooden tools; campfires provide a social focus; and the cooking of food not only renders it more digestible but kills parasites. Yet it's hard to know when fire became a regular feature of hominid life or exactly when the cognitive apparatus had developed that allowed the domestication of this phenomenon, so frightening in its natural, uncontrolled form. Even today, many campfires are of the kind that would leave little lasting evidence behind; and hearths such as those reported from Terra Amata are rare indeed in the archaeological record before about 150 kyr ago. One swallow doesn't make a summer; and any innovation has to become a regular feature of life before its full impact is realized. Terra Amata considerably predates the period within which hearths are commonly found at archaeological sites, and what one should make of this is not at all clear. The advantages of domesticating fire are so great that, once technologies for its control had been developed, it's hard to see why they should not have spread rapidly.

Fire and shelter aside, though, localities like Terra Amata and Arago leave one with the impression that, in terms of their subsistence and the tenor of their daily lives, their occupants had done little more than to refine the routine of their ancient precursors. This echoes the pattern of change that seems to have applied throughout human evolution up to this point: economic innovations were sporadic and incremental, and largely independent of the coming and going of species. There is little doubt that *Homo heidelbergensis* subsisted on both animal and vegetal resources, lived in relatively small, mobile groups, and sought shelter from the elements. Its stoneworking technology was certainly no more sophisticated than that of later *Homo ergaster*; indeed, the last major technological innovation (by what species is uncertain) had occurred in Africa 600 kyr before Terra Amata, with the introduction of a means of making flatter, thinner hand axes: tools more sophisticated than any at the early European sites. How individuals of this time interacted and viewed the world remains a closed book; the lens of archaeology is just not precise enough. It may, nonetheless, be telling that it discloses no convincing evidence of symbolic activities among *Homo heidelbergensis*—although it appears that stone tools had been used to deflesh the Bodo skull, for reasons that remain entirely obscure.

Just possibly, though, there is evidence in the anatomy of this species that bears upon how individuals communicated. For it is with *Homo heidelbergensis* that the element of human language potentially enters the picture (although it has occasionally been argued, speciously in my view, that language in some recognizable form had been around much earlier—even among the very earliest species of *Homo*). In *Homo heidelbergensis*, for the first time, we encounter a skull base anatomy that suggests its possessors had all the peripheral equipment necessary for speech production. We will explore language and speech later in this chapter; suffice it to say for the moment that while speech is made possible by the specialization of structures in the throat that may imprint their presence on the bottom of the skull, language is a product of the brain. And while raw brain size in *Homo heidelbergensis* was quite impressive, the archaeological record left by this species provides no evidence of the complexity of behavior that seems to be so integrally associated with the language capacity. Thus in all probability *Homo heidelbergensis* did not possess language in any familiar sense, although it's probable that, however individuals of this species communicated, they did so in quite sophisticated ways.

## *The Neanderthals*

*Homo heidelbergensis*—or, more probably, a species not too unlike it, minus some specialized features such as the development of enormous sinuses inside the skull—was quite plausibly our ancestor. It may also have been the ancestor of *Homo neanderthalensis,* a specialized and recently extinct human relative. The Neanderthals, as humans of this kind are more commonly known, are unquestionably the best documented of all the many extinct kinds of hominid; and particularly since it shared Earth with *Homo sapiens* over an extended period of time, *Homo neanderthalensis* provides us with the best yardstick we have for assessing the degree of our own uniqueness. It is thus particularly important to understand the Neanderthals and their achievements if we are to appreciate the gap that still yawns between ourselves and our closest relatives in nature.

Exactly when the Neanderthal story begins is tricky to pinpoint, largely because the fossil record is a little fuzzy. Some exquisitely preserved 300-kyr-old human fossils, from the "Pit of the Bones" in the Atapuerca Hills of northern Spain, are said to show various incipiently Neanderthal characteristics. However, they most likely represent an as yet unnamed species that was related to the Neanderthals but was by no means the same thing. What's more, we know rather little about these hominids other than their anatomy; they are unaccompanied by artifacts of any kind. It may be that we can see a "humanizing" feature in grooves between their teeth which show the Atapuerca people used toothpicks; but in fact this is something that first shows up in isolated teeth from Ethiopia that are well over 2 myr old, and a chimpanzee with nasal congestion has been seen attempting to unblock his nose with a stick.

All fossils plausibly viewed as Neanderthals are considerably younger than the Atapuerca folk, broken crania from the German site of Ehringsdorf being the oldest contenders at (probably) something over 200 kyr in age. There is more evidence from the long period of intense cold around the height of the next-to-last major glacial (around 180 to 150 kyr ago); but it is only in the last interglacial (around 120 kyr ago) that the record really begins to pick up. From that point on, literally hundreds of sites in Europe and western Asia, from the Atlantic to Uzbekistan and from Wales to the Mediterranean, have yielded evidence of Neanderthal occupation before the species finally disappeared at about 30 kyr ago.

Physically, the Neanderthals were quite striking. They had brains that were as large as our own but that were differently shaped. Although a recent analysis has concluded that Neanderthal endocasts reflect a substantially modern level of external brain organization and asymmetry, it is perhaps significant that the Neanderthal frontal association cortex (where, as you'll recall, a lot of our "thinking" is done) was rather constricted and flattened, like that of earlier hominids and in distinct contrast to the upward expansion of this area in ourselves. Unlike our high and rounded cranial vault, that of the Neanderthals is long and low, reflecting the distinctive brain shape of the species; and it protrudes curiously at the back. At the front, a pair of large, rounded, and continuous ridges overhangs the eyes, and the face protrudes markedly, particularly in the midline, where the cheekbones sweep sharply back from a very large nasal aperture wherein reside some very unusual structures. Modern humans, in contrast, are mostly smooth browed (and where there are ridges above the eyes, they are of a different shape); and the small face is tucked below the front of the skull. The size and disposition of the upper face reflects itself in the structure of the lower jaw. Where modern humans possess a jutting chin and short tooth rows, the lower jaw of Neanderthals is vertical or receding in profile, and the tooth rows are long. The skulls of the two species also differ in numerous finer points of detail.

As for the body skeleton, the Neanderthals were in some respects much more robustly built than we are, and on the whole they were shorter in stature, though some were quite tall. Limb proportions were a little different from ours, too, with relatively short upper arm and leg segments. The main joints of the Neanderthal body are larger than ours are, and the arm and leg bones are thick walled, although their rather unobtrusive muscle attachment surfaces belie the strong build of the bones. The structure of the shoulder blades bespeaks a very powerful upper arm musculature, and certain details of the construction of the massive Neanderthal pelvis hint at a minor difference in gait between these extinct humans and ourselves. Since the development of bones is at least partly dependent on the muscular forces acting upon them, it has been suggested that Neanderthal robustness may have been due to a particularly demanding physical lifestyle; but in view of the fact that many of the anatomical peculiarities of the adults show up even in very young Neanderthals, this seems at the very least rather dubious.

The Neanderthals lived (as a species) for a long time, in a vast area, and during a period that was subject to quite violent climatic swings. It's thus obvious that there was no "typical" Neanderthal environment and that these hominids must have been quite adaptable in the way in which they exploited their varied habitats. On balance, though, those habitats were rich ones. Many Neanderthals shared the landscape with a multifarious fauna of mammals of a huge range of body sizes, all the way from shrews to mammoths and including many relatively vulnerable herding animals of moderate body size, such as reindeer and wild cattle. We'll see in a moment that there is considerable argument over how skillfully the Neanderthals exploited those potential resources; but it is clear that their way of life was a successful one that ultimately succumbed only to a new and unprecedented presence on the landscape: ourselves. It was not an easy existence, though; far from it. For the skeletal remains of individual Neanderthals bear witness to short and difficult lives. Few Neanderthals survived to the age of forty, and Neanderthal skeletons routinely bear the marks both of trauma and of dietary stress.

European *Homo heidelbergensis*, as we've seen, was typically equipped with a rather crude stone tool kit, although the Acheulean was certainly established in Europe by 300 kyr ago. At some point over about 200 kyr ago, however (dating is hazy), a radically new kind of stone toolmaking entered the record, announcing the debut of the Middle Paleolithic (in Africa, the Middle Stone Age). This prepared-core technique involved carefully shaping a rock nucleus, to the point where a single blow would detach a large flake that needed only minor modification to become a finished tool. Hammers used to make such tools were usually "soft," that is, made of bone or antler rather than of stone; or an intermediate soft "punch" was employed between the hammer and core. The advantage of this approach was that it gave each tool a long, continuous cutting edge: something that simply reducing a core to a particular shape with a long succession of blows did not provide. This indirect approach to tool production also bespeaks a sophisticated insight into both the properties of the stone being worked and how they could be employed to best advantage. Here, then, was another major conceptual leap in stone toolmaking, although it's impossible to say by which kind of early human this leap was made. But whoever may have been responsible for its invention, it is clear that this new kind of toolmaking was most fully and expertly exploited by the Neanderthals.

## *Neanderthal Life*

Although the tools associated with the very earliest Neanderthals are of a fairly generalized Middle Paleolithic type, this tradition rapidly blossomed into the Mousterian industry that has become firmly (though as we'll see not exclusively) identified with the Neanderthals over the entire broad swath of Europe and western Asia in which their fossils have been found. It's interesting in this context that the Mousterian, or something virtually indistinguishable from it, has also been found in northern Africa, whence no Neanderthals are known. Mousterian stone tools are beautifully made and represent variations on a variety of themes: an early classification recognized over eighty variants (small hand axes made on flakes, scrapers, points, and so forth), although it has since been recognized that many allegedly distinct "tool types" actually represent different stages in the resharpening of a more limited number of basic tool forms. This pattern of use and reuse (particularly of tools made from desirable raw materials that were hard to obtain) may be one reason why it has been so difficult to develop a basic typology of the tools found at Mousterian sites. And while the argument has been made with some justice that Mousterian tools and tool kits are essentially the same wherever they have been found in time and space, there are some indications that diverging regional toolmaking traditions developed in particular areas.

If, then, Mousterian craftsmen were indeed aiming to produce tools that conformed to a variety of basic patterns they held in their minds, and were capable of at least some degree of local innovation, we are witnessing yet another cognitive refinement among these humans of the Middle Paleolithic. We have no way of assessing exactly how this particular refinement was reflected in their broader being; but although in comparison to their successors these Mousterian toolmakers were using relatively simple techniques to make a limited variety of products, they showed a then-unprecedented appreciation of what could be done with stone.

This makes it all the more surprising that these accomplished stone workers appear hardly ever to have employed bone and antler as raw materials in toolmaking. This contrasts very distinctly, as we saw in the first chapter, with the Upper Paleolithic people who replaced them. These later people not only used and decorated such materials, but they exploited them extremely imaginatively in a host of uses. From the study of the wear on Mousterian stone tools, it is evident that some were used in working wood (sharpening spears?) and in scraping hides; but as far as the basic tool kit was concerned, stone was the material of choice. This surely hints at some kind of intellectual limitation in the Neanderthals: they had inherited the concept of working stone and had taken this practice to previously unheard-of levels of refinement, but they never got the idea of moving beyond this. If the Neanderthals did sharpen spears, this practice had been bequeathed to them by remoter forebears: a wooden spear point possibly as much as 350 kyr old has miraculously been preserved at the English site of Clacton-on-Sea, which amply predates any Neanderthal locality. Spears can, of course, be used for throwing as well as for thrusting; but it is debated whether the Neanderthals ever hafted stone points as missile tips (or for other uses). Impact damage on a few Mousterian stone points seems to indicate that at least occasionally Neanderthals attached such items to the ends of spears; but such evidence is rare at best, and it seems pretty certain that stone-tipped spears, at least, were not a part of the regular Neanderthal arsenal.

How effective Neanderthals were as hunters is also debated.

If Neanderthals had wooden spears, which we can be pretty confident about, then their obvious use was in hunting; but thrusting spears are only useful at very close range, which is highly dangerous (as injuries to Neanderthal skeletons may well attest); and we have no good evidence, even inferential, that Neanderthals used thrown spears. Better evidence for hunting activities comes from the nature of the animal bones found at certain Mousterian sites, although once again we face the difficulty of discriminating the results of scavenging from those of hunting. At some Mousterian localities, an unexpectedly high proportion of bones represent the meaty parts of medium-bodied mammals, such as horses and reindeer. This might be taken as evidence for hunting rather than the scavenging of carnivore kills; but evidence of this kind is sparser than we might wish, and it's possible that the Neanderthals involved were simply selective in what they scavenged. Such selectivity might also account for the preponderance of particular animal species at Mousterian sites, a not uncommon finding. In a few places, Mousterian tools have been found in association with the remains of large mammals lying at the bottom of cliffs, and this has been advanced as evidence of "fall-hunting," in which the victims were driven over the cliffs to their deaths. Such cases are rare, however, and more mundane explanations have been offered for the bone accumulations.

The question is, however, far from simple. At one Italian site, there is evidence for Mousterian occupation at two distinct periods. During the first of these, which fell in the last interglacial (about 120 kyr ago), occupations appear to have been quite brief, and animal remains were preponderantly the heads of older individuals who might have died naturally. This looks very much like evidence for Neanderthal scavenging, rather than hunting. At the later, cooler time, about 50 kyr ago, the animal remains mostly came from individuals in the prime of life and consisted of bits from all parts of the body. Together with larger numbers of stone tools, this pattern seems to suggest longer occupations and the use of ambush hunting. Whether this difference was due to an improvement of techniques over time or simply to differing responses as the climate cooled—or to something else—is anybody's bet. But at the later time, at least, the nature of the animal-bone assemblage does suggest active hunting by Neanderthals of medium-sized herbivores, such as red and fallow deer.

Most archaeologists would, in fact, concede that Neanderthals at least occasionally hunted mammals of this size, although they would rarely have tackled anything much bigger. Humans in general are (and were) slow-moving creatures, and modern humans are incomparably successful hunters because they exercise craft, guile, and an unparalleled perception of the cues offered by the outside world. It is not at all apparent that Neanderthals were capable of similar skills; and if they weren't, they would have been inherently limited in the kinds of hunting they were able to carry out. Probably hunting by guile as we know it is a peculiar property of our own species. But hunting is, of course, very far from the whole economic story. Let's not forget that in all periods the Neanderthals almost certainly gained most of their sustenance from plant foods they gathered.

Our view of Neanderthal life is probably biased considerably by the fact that most of their sites excavated thus far have been located in cave mouths and rock shelters. These may well have been preferred places, not just because they offered protection from the elements, but because they are often located near good sources of flint, the ideal raw material for stone toolmaking. Open-air sites, more susceptible to destruction by natural forces and agricultural activities, have attracted less attention; but as more effort is directed toward locating and excavating them, it is possible that our perspective on Neanderthal life will shift a little. What we know thus far suggests that Neanderthals often revisited campsites at particularly favored spots and remained there for varying periods of time. Hearths are a regular feature of Mousterian sites, and occasionally postholes (and at one site a natural cast of a tent peg) have provided evidence that from time to time Neanderthals rendered their camping sites more comfortable by the rigging up of shelters. Indeed, there is a very recent report of a Neanderthal structure quite deep within a cave, indirectly suggesting that some form of artificial lighting was available.

In comparison with those of their Upper Paleolithic successors, however, Neanderthal sites are pretty simple and unstructured (and suggestive of small group sizes), though there may be a couple of exceptions. Whereas at most Mousterian sites tools and bones are relatively randomly scattered around, at the Israeli locality of Kebara hearths are concentrated in the center of occupied areas largely cleared of rubble, and flints were knapped and used next to the fires. Animal bones, on the other hand, appear to have been flung toward the back of the cave. This hints at some basic pattern of organization of the living space. Even more intriguing is a recent reappraisal of the site of Combe Grenal in France (whence the tent-peg cast came), in which it is argued that males and females occupied distinct areas of the site, distinguished by different patterns of tools, bones, and fire use that reflect distinct types of economic activity. The females, according to this reconstruction, foraged in the local area for plant foods and cooked over low flames, living apart from the males. The latter, in contrast, ranged more widely in search of animal prey and may have returned home only occasionally, bringing with them marrow bones that had to be heated to higher temperatures to extract their nutritional content. Such total separation of male and female economic activities—even lives—is radically different from anything known among modern hunters and gatherers, and it will be interesting to see whether or not this interpretation is sustained by future studies.

In terms of overall economic strategy, it has been suggested that a major distinction between Neanderthals and modern hunters and gatherers is that the former were foragers, while the latter were and are collectors. The distinction is this: Foragers roam around the landscape in an opportunistic manner, making use of any resources they encounter. Collectors, on the other hand, plan the use of resources whose whereabouts are known and carefully monitored. Forward planning is a conspicuous characteristic of human activity, and its absence or lesser expression among Neanderthals would certainly speak volumes about relative cognitive capacity in *Homo neanderthalensis* and *Homo sapiens*. Another intriguing suggestion of possible differences between these two species in their approach to exploiting their environments comes from studies of sites in Israel. A major puzzle in human evolution has been posed by the fact that whereas the arrival in Europe of modern humans heralded the quite abrupt extinction of the Neanderthals, the two kinds of humans managed to share the Levant for an extended period, from about 100 to 40 kyr ago. Sites yielding the fossils of both are known from Israel in this interval.

It has been suggested that the Neanderthals were "cold-adapted" and were thus able to maintain or reestablish occupation of the region during cooler periods, only ceding possession to the tropically derived *Homo sapiens* during warmer times. Perhaps this was so; but what is most intriguing is that throughout this time in Israel, both species wielded virtually identical stone tool kits: the local version of the Mousterian. The sites left behind by both are also similar in character; but a meticulous analysis of faunal remains and the patterns of breakage and wear on stone tools at three Levantine sites—two occupied by Neanderthals, the other by early moderns—has hinted at a difference in way of life.

The analysis started from the notion that the early Levantines had two basic economic strategies available to them. One of these, the "circulating mobility" pattern, involves the regular shifting of campsites as each group moves around its territory en masse, responding to seasonal change in available resources and to the routine depletion of resources that occurs around any campsite. The other strategy, "radiating mobility," places a semi-permanent campsite somewhere in the middle of the territory, with smaller satellite camps in outlying areas near particular resources. Satellite camps would normally be visited not by the whole group at once, but for short periods by specialized parties, with a return to the central base at the end of each foray. Groups employing radiating mobility seek to maximize the length of time they reside at base, until the length of each excursion becomes too extended to be worthwhile. As a result of finely detailed analysis of the Neanderthal and early modern sites, the researchers concluded that the former had practiced circulating mobility, while the moderns had preferred the alternative. They also concluded that ecological differences were insufficient to explain these results. Of course, the sample of sites was small, but these findings are enough to suggest that Neanderthals and early moderns in the Levant exploited their environments in different ways: ways that lend support to the notion that Neanderthals were foragers and the early moderns collectors.

The indirect evidence of the tools and sites thus hints at an innate difference in ways of viewing the world between Neanderthals and modern humans; and we'll discuss later why early moderns in the Levant left no evidence of the striking technological innovations and symbolic behaviors that characterized their later counterparts in Europe. Meanwhile, in order to put these developments in perspective, it is worthwhile to look at the evidence for symbolic behaviors among Neanderthals.

## *Neanderthals and Burial*

Early studies dramatically associated Neanderthals with bizarre rituals involving "bear cults" and cannibalism; but more recent reappraisals of the alleged evidence for such behaviors have resulted in very much more cautious conclusions. However, the Neanderthals did indulge in one very specific behavior that certainly hints at such qualities as awareness of death and possibly also at spiritual belief. This is burial of the dead. Actually, the idea that Neanderthals ever practiced burial has recently been disputed, but there's nonetheless no doubt in my mind that, at least occasionally and very simply, the Neanderthals did bury their dead, laying them out mainly in flexed postures—which might, of course, have been purely for the practical reason that it required digging a smaller pit. If so, it's quite possible that burial of the dead may not have had the same symbolic significance to the Neanderthals as it does to us: it might simply have been one way of avoiding a particularly unpleasant form of clutter in the living space or of discouraging the incursions into living sites of hyenas and other scavengers.

What's more, it's far from clear that Neanderthal burial ever involved the sort of ritual that is invariably associated with this practice among modern human populations—and which, as we saw in the first chapter, is so dramatically evident in many burials of the Upper Paleolithic. The strongest claim for elaborate Neanderthal burial comes from Shanidar cave, in Iraq. There, a 50-kyr-old grave was found to contain an unexpectedly high level of pollen. This suggested to its excavators that the deceased had been laid to rest on a bed of spring flowers. There are, however, other ways in which the pollen might have entered the grave, although the occurrence is certainly suggestive. Possibly more so is another burial from the same site, where the grave contained the remains of an old male who had for years been crippled by a withered arm. The world of the Neanderthals was a harsh one, and such a handicap would almost inevitably have been rapidly fatal if the individual concerned had not enjoyed the long-term support of his group. The best evidence we can hope to find for burial ritual, however, is the presence in the grave of so-called grave goods: artifacts, food, and so forth that the deceased might find useful in a future life. Upper Paleolithic burials are typically rich in such goods, but all such items that have been reported from Neanderthal burials are the kinds of things that might well have found their way into the grave by accident. Stone tools and animal bones lie around in profusion on the floors of Neanderthal sites; and it would be surprising indeed if a few such objects had not coincidentally found their way into graves while they were being filled.

Ironically, two Mousterian burials do appear to show convincing evidence for the interment of the deceased with grave goods; but they are in early Levantine sites—Skhul and Qafzeh, both around 100 kyr old—that were left by early modern humans (or, in the former case, something very like them and certainly not Neanderthals): people who looked pretty much like you and me. Another tantalizing hint of cognitive differences between Neanderthals and moderns? These grave goods are hardly very impressive, though. At Skhul, the lower jaw of a wild boar was found "clasped in the hands" of an effectively modern human; and at Qafzeh a large pair of fallow deer antlers overlay a modern corpse. These cases apart, there are no unambiguous grave goods associated with any Mousterian burial—Neanderthal or otherwise. Of course, there are very much easier ways of disposing of bodies than by digging graves for them, and it is still hard to avoid the conclusion that the mere act of burial tells us something about the Neanderthals as sentient beings. It must, for instance, indicate at the very least a strength of attachment between individuals that transcends anything seen previously: a gesture toward the buried that was far from obligatory for any but emotional reasons. However this may be, though, it is difficult to sustain the notion that Neanderthal burial represented symbolic activity, as opposed to the simple expression of grief and loss.

Well, if burial is no proof of symbolic activity, what else might we look for? Our only other avenue at present is to look directly for material expression of the symbols themselves—for, after all, art, symbol, and notation are among the most striking components of the cultural heritage left to us by the Upper Paleolithic peoples of Europe. The earliest putative examples of symbolic activity actually predate the Neanderthals. A fragment of bone from Bilzingsleben in Germany bears a curious set of incisions that hardly qualify as art but may have been the deliberate work of a human hand some 350 kyr ago. The pre-Mousterian site of Berekat Ram in Israel, about 230 kyr old, has produced what is claimed to be a rough outline of a female human engraved on a pebble. These early manifestations—neither of them very impressive and neither unambiguously deliberate—must be regarded as oddities, whatever they may reflect of the sentience that lay behind them. What is more significant is that nothing in the way of putative symbolic production associated with the Neanderthals is much more striking than these earlier examples —with a single exception we'll come to in a moment. Within the documented Neanderthal span, we very occasionally find a bone bearing incision marks, a tooth or bone fragment with a hole battered through it, or a piece of stone exhibiting strange hollows. Once in a while at a Neanderthal site we find a fossil, imported from somewhere else—which at least indicates a sort of aesthetic curiosity—or deposits of ocher that might or might not have been used in body painting. But these meager expressions are about it, even where they may have been intentional— and some kinds of butchery marks on bone, for instance, can easily be mistaken for deliberate series of incisions. The very rarity of occurrences such as these indicates that symbolic expression was not a regular aspect of Neanderthal life, and we are forced to conclude that symbolic behaviors—or their material expression, at least—were not a significant part of the shared cultural existence of these early humans. The contrast with the symbolic outpourings of the Cro-Magnons could hardly be more dramatic.

## *The End of the Neanderthals*

In the thirteen thousand years following about 40 kyr ago, the Neanderthals were edged out of their last refuge in western Europe and into extinction. As we've seen, the early modern humans who displaced them were equipped with the full cultural and technological panoply of the Upper Paleolithic. This was not an overnight event, though, even in specific places. At a few sites in western France and northern Spain, we find evidence of a material culture known as the Châtelperronian, in the period between about 36 and 32 kyr ago. This culture incorporates certain aspects of the Upper Paleolithic (in particular, the production of multiple long, thin "blade" tools, notably burins, and a certain amount of work on bone), but it is now generally agreed to have been the work of Neanderthals. At Roc de Combe and La Piage in western France, the Châtelperronian alternates in the archaeological layers with the Aurignacian of the earliest moderns, showing that the replacement of the Neanderthals by the moderns was not a one-step process. In a Châtelperronian layer at Arcy-sur-Cure (along with a Neanderthal fossil) was found a carved bone pendant, complex in shape and quite finely made and polished, that was most likely an item of personal adornment worn by a Neanderthal.

This unique object was, however, made at a time when modern human incursion into the area had already been achieved; and increasing numbers of archaeologists are taking the view that certain aspects of the Châtelperronian were the result of the Neanderthals' copying, either directly or indirectly, of Upper Paleolithic technology. An alternative suggestion is that such objects might have been acquired by Neanderthals through trade with modern humans. Either way, they were not a Neanderthal invention. Whatever the case, it appears that the Châtelperronian was a very late terminal development out of the Mousterian. The Middle Paleolithic did not transform itself into the Upper Paleolithic via the Châtelperronian, and neither did the physically very different Neanderthals evolve into modern humans.

One thing the Châtelperronian certainly does not do is to blur the distinction between Neanderthals and modern humans, either technologically or symbolically. Almost everything we know about the Neanderthals points to the conclusion that here were beings, big brained though they might have been, who were cognitively distinct from ourselves. And it was this difference, almost certainly, that accounts for their disappearance. Modern humans arrived in Europe about 40 kyr ago, and by about 27 kyr ago the Neanderthals were gone forever. Admirable the Neanderthals certainly were in coping successfully for so long with an erratic and often severe environment. But what they could not cope with was an entirely new factor: us. It will always be debated whether the Neanderthals became extinct because they were inferior competitors for resources or because they were physically wiped out by invading modern humans. Probably the truth lies somewhere in between. But the disappearance of the Neanderthals is not in itself the extraordinary phenomenon it is often made out to be. After all, almost all species that have ever lived have become extinct, and those living today are just those that have yet to suffer this fate—even as the overall exuberant diversity of living forms continues.

It is, of course, profoundly misleading to see the Neanderthals simply as an inferior version of *Homo sapiens*, for there are many ways to play the evolutionary game, and the Neanderthals played to a different set of rules. With changing circumstances, their rules eventually let them down—as ultimately could ours. Nonetheless, it is in comparing ourselves to the Neanderthals that we can most truly gauge our uniqueness in nature. We have, of course, to concede that the archaeological record is but a dim reflection of the full behavioral panoply of any hominid species, as the example of the Tasmanians attests in our own case. But any difference as dramatic as that evident between the records bequeathed to us by the Neanderthals and the first modern Europeans is clearly significant. Nonetheless, it remains true that archaeology will almost certainly never be able to answer directly one of the most pivotal questions about the Neanderthals: Did they have language or articulate speech? Here we have to return to anatomy.

## *The Evolution of Speech and Language*

Among all our remarkable attributes, the most striking is our possession of language. And knowing whether or not this is something that was shared with the Neanderthals is crucial to the assessment of the size of the cognitive gap between them and us. For, if they had language, they *were* us in a profound sense, despite their many physical differences and the limited range of their material productions. The problem is that language is essentially a product of the brain, and, as we've seen, the evidence of fossil brain casts is hard to interpret. As far as can be told, some increase in brain complexity accompanied the long process of brain enlargement; but especially in view of the apparent diffusion in the cortex of various functions related to language skills, it's tough to be precise about what this means. We cannot, for instance, be confident in drawing a causal correlation between hemispheric asymmetries and language. But while language is an ephemeral behavioral attribute, its companion, articulate speech, is a rather different matter. The sounds necessary for language are produced outside the brain, by the structures of the vocal tract. And while the vocal tract itself does not preserve in fossils, its roof does, in the form of the base of the skull.

The principal structures of the vocal tract are the larynx, the structure in the neck that contains the vocal folds (cords); the pharynx, a tube that rises above it and opens into the oral and nasal cavities; and the tongue and related structures of the mouth. Basic sounds are generated in the column of air that rises from the lungs, through the action of the larynx. There is then, however, the possibility of further modification of those sounds in the air column above. In typical mammals, including apes and all other nonhuman primates, the larynx lies high in the neck, and the pharynx is short. This shortness severely limits the ability of the surrounding musculature to manipulate the pharynx and thereby to modify the vibration of the air passing upward from the larynx. In adult humans, in sharp contrast, the larynx lies low in the neck, lengthening the pharynx and making further sound modification possible. On the minus side, this anatomical arrangement also impedes the ability to breathe and swallow simultaneously—which is why, regrettably, almost every one of us knows of someone who choked to death. Interestingly, human infants—who need to be able both to breathe and to swallow during long bouts of suckling—are born with the primitive mammalian vocal tract anatomy. The larynx only later descends to the low position in the throat, beginning at about two years of age.

The consequences of these anatomical shapes for speech production are obvious. The sounds that make articulate speech possible are produced largely by muscular modification of the pharynx's cross-section and can only be produced by a long, high pharynx of adult human type. Apes can't make them; babies apparently can't, either. And we see this limitation reflected in the shapes of their skull bases. Apes of all ages and human infants have typically mammalian flat skull bases. Modern human adults, on the other hand, show strong downward flexion of the bottom of the skull, to accommodate the tall, looping pharynx. The difference is truly striking. And the only *obvious* advantage we can ascribe to skull base anatomy of the adult modern human type —or at least the only one we can envisage by reference to a living model—is that it confers the capacity (or at least the potential) for speech. Is the form of the skull base, then, a magic indicator of speech capabilities? Well, as always, it's not quite as simple as that.

We know that apelike morphology precludes speech of our kind, and we also know that fully modern human morphology permits it. But we can't be sure exactly what any intermediate stages preserved in the fossil record might imply. We do know that all australopiths have more or less flat skull bases, which suggests that these creatures did not have a mechanism to permit much if any more complex sound production than we find in apes. So far so good; but moving up in time we enter a gray area, not simply because it is difficult to envision what a slight change in skull flexion toward the modern human form means in terms of the specifics of communication, but because few studies have been done. Some increase in cranial base flexion, suggesting a small degree of laryngeal descent, is seen in *Homo ergaster* and *Homo erectus*. The dashing Jeffrey Laitman (his adjective)—a leader in this field of investigation and a professor at Mount Sinai School of Medicine in New York City—is inclined to believe that at least initially this descent was probably due to respiratory rather than vocal modifications; but, whatever the case, this change set the stage for an enlarged vocal capacity. Cranial base flexion to a virtually modern degree has been detected in individuals of *Homo heidelbergensis;* but the neat picture of gradual modification in the peripheral vocal apparatus more or less in step with overall brain-size increase is apparently sullied by the Neanderthals. The one Neanderthal skull—a relatively late one from France—that has been looked at in detail from this point of view is said to have only a modestly flexed cranial base, even after recent reconstruction. What's more, computer simulations of the sound-producing potential of its vocal tract suggested that in life this individual was incapable of producing certain sounds that are essential to modern speech. Puzzlingly, though, a brief appraisal of an earlier Neanderthal skull from Italy revealed a greater degree of skull-base flexion, implying a better vocal potential. The question of Neanderthal speech-producing potential has yet to be settled.

More detailed and extensive studies along this line certainly hold the promise of increasing our understanding of the evolution of (potentially) speech-producing capabilities in humans. Unfortunately, though, the researchers involved appear to have been discouraged by early criticism, and this promise has been far from fully exploited. Thus the conundrum of the Neanderthals remains, among many others. We'll return to this point in the next chapter.

Happily, though, the skull base is not the only avenue available for investigating the question of speech capability. An unexpected clue turned up, for example, during the study of the Turkana boy *Homo ergaster* specimen. Researchers noticed that in this individual, the thoracic vertebral canal—the upper part of the bony tube through which the spinal cord descends to control the various regions of the body—resembles that of modern humans less than do the parts of the canal that lie below it. The thoracic canal houses the portion of the spinal cord that conducts instructions from the brain to the muscles of the rib cage and abdominal wall. These muscles, in turn, provide precise control of the way in which air is expelled from the lungs. The difference in the size and shape of the thoracic canal in *Homo ergaster* and *Homo sapiens* suggests that in *Homo ergaster* the nerve supply to this musculature was less well developed. And this in turn implies that *Homo ergaster* possessed less precise control than we do of the rib cage, hence of the air column rising from the lungs. It is hard to see why any difference in the thoracic canal should exist, apart from the added demands placed on the respiratory system by speech; and if *Homo ergaster* thus lacked thoracic mechanisms necessary in the sound-modulation process, we have independent evidence to support Laitman's conclusion that incipient flexion of the cranial base in *Homo ergaster* was not associated with producing the sounds of speech.

One other factor has been held out as a potential indicator of the speech capabilities of early humans. The hyoid, a free-floating structure in the throat, serves as a point of attachment for a variety of vocal tract muscles. And although small and delicate, the bony portion of the hyoid is occasionally preserved in fossil form. Until recently no fossil human hyoids were known, but lately one has turned up. The hyoid in question belongs to the Neanderthal individual buried at Kebara, in Israel, and its discovery immediately raised hopes of illuminating, if not solving, the enigma of Neanderthal speech capabilities. But that, of course, would be too good to be true. As you might expect after reading the last few pages, the analysis of the Kebara hyoid has merely succeeded in unleashing an appropriately vociferous argument about its significance. The problem is that, although this bone looks pretty much like its counterpart in modern humans, the preserved bony element forms only a limited portion of the complete hyoid structure. And what the long-disappeared cartilaginous part of the Kebara hyoid looked like is anyone's guess. The issue remains unresolved.

Physical evidence for the evolution of human speech capabilities thus remains equivocal. What's worse, to the extent that it is interpretable at all, this evidence coexists uneasily at best with what is normally inferred about linguistic behaviors from the archaeological record. It is very hard to avoid the conclusion that articulate language is quite intimately tied up with all of the other mysterious and often unfathomable aspects of modern human behavior. Yet we know that effectively modern skull-base anatomy appeared long before we have any convincing archaeological evidence for complex symbolic behaviors. Indeed, as we've seen, the inferred form of the vocal tract in *Homo heidelbergensis* is more or less modern, in which case acquisition of the vocal apparatus permitting speech preceded the arrival of *Homo sapiens* by several hundred thousand years. Perhaps, then, the descent of the larynx not only began as a function of some sort of respiratory modification, as Laitman suggests, but also continued for nonlinguistic reasons. The simple fact that we have so far been unable to come up with any clearly demonstrable alternative to speech as an evolutionary advantage for laryngeal descent is not necessarily reason for thinking there *can* have been none. And if there *was* such an alternative, it implies that the evolution of the vocal tract and of those centers of the brain concerned with language did not go hand in hand, which in turn makes reconciling the anatomical and archaeological evidence less problematic.

Simultaneous acquisition of both the central and the peripheral apparatuses necessary for language would have been quite a developmental trick for evolution to pull off, and a multistage process is certainly easier to envisage in both developmental and evolutionary terms. If a vocal tract capable of extensive sound modification had become established for nonlinguistic reasons, it could have remained around for a protracted time as what we can in retrospect view as an exaptation for speech: a necessary structure that was initially acquired in another context. There's nothing very unusual about that, for we've seen that in the course of evolution, new structures routinely arise for functions other than those for which they are eventually conscripted. All that would have been needed in this case was *some* countervailing advantage against the drawback of choking. Such an advantage might even have lain in one form or another of improved vocal communication—just not in language as we are familiar with it. Under an evolutionary scenario of this unexceptional kind, the *full* potential of this new anatomical configuration would only have been realized once the appropriate brain mechanisms had been acquired. And these neural mechanisms would, of course, only have worked in practice because the necessary peripheral exaptation already existed.

All of which brings us back to the question of whether Neanderthals had language. To which the answer is almost certainly no, at least in the form in which we are familiar with it. The peripheral apparatus necessary for speech had been around before the Neanderthals, but it has been argued that in the apparently flattish cranial bases of at least some of these hominids, we see an unprecedented reversal of laryngeal descent. If so, it's quite likely to have been a matter of respiratory accommodation to some environmental necessity, unrelated to speech as such; but in any event, if we combine this with the absence of any substantive archaeological evidence of symbolic behavior, it seems reasonable to conclude that the Neanderthals did not communicate as we do. What's more, in the first chapter of this book we saw that there is abundant archaeological evidence of the astonishing accomplishment of the first modern humans who occupied Europe, displacing the Neanderthals in the process. Art, symbol, music, notation, language, feelings of mystery, mastery of diverse materials, and sheer cleverness: all these attributes, and more, were foreign to the Neanderthals and are native to us. We are the product of a long evolutionary story, peopled by many different actors, and in our fossil record we can see the erratic accretion, over a vast span of time, of various characteristics we recognize as part of the complex of anatomies and behaviors that distinguish us from our closest living relatives. But in everything we can reasonably infer of the behaviors of our close relatives the Neanderthals, we still find the yawning cognitive gulf between modern *Homo sapiens* and the rest of nature only partly bridged. What we know of the Neanderthals narrows that gap, it's true; but they and their forebears only half closed it at best. The first bipeds and stone toolmakers possibly apart, it is only with the arrival of *Homo sapiens* that we find true innovation: a radical departure from the pattern of sporadic improvement on existing themes that had characterized the rest of human evolution.

## *The Emergence of Modern Humans*

I have already described in chapter 1 the remarkable achievement of the first modern humans who entered Europe. Where these people entered it *from* is far from clear, and I'll return to this question shortly. For the moment, though, it's enough to point out that the first modern Europeans arrived fully fledged from elsewhere; this is most certainly not the place where modern humans *originated*. Where that place of origin was is complicated by the fact that humans may be "modern" in two distinct senses: anatomical and behavioral, as we'll see; but there's little doubt that *Homo sapiens*, as recognized from fossils, arose in Africa or nearby. Molecular studies also suggest an African origin for our species, perhaps 130 to 160 kyr ago. And while that inferred date is subject to some dispute, the fact that African populations show more molecular variability than those from other continents shows pretty convincingly that *Homo sapiens* has been differentiating there considerably longer than elsewhere.

The fossil record, sparse as it is, tells a similar story. Perhaps as much as 120 kyr ago, some fairly modern-looking humans camped near the southern tip of Africa, at the sites of Klasies River Mouth. They made tools of Middle Stone Age type (more or less equivalent to the Mousterian of Europe and western Asia); but there are hints that the way they used their living space was more modern in style than their artifacts, even though they left behind no material evidence of symbolic activities. The human fossils from Klasies are few and fragmentary; but, interestingly, they have been interpreted as the remains of a cannibal feast: the earliest substantial evidence we have of such behavior.

Other sites in Africa have also yielded early humans of modern anatomy, but are bedeviled by uncertainties of dating. A fragmentary modern cranium from Omo, in Ethiopia, may be as much as 125 kyr old, although other African specimens in this time range retain a more archaic form. In terms of its fauna, the Levantine region was effectively part of Africa, and it is interesting that it is from this area that the only substantial and well-dated fossil evidence for very early modern humans comes. In Israel the sites of Skhul and Jebel Qafzeh have recently been dated to close to 100 kyr (the first a little more, the second a little less). As we've seen, these sites both yielded tools of Mousterian type, but each did suggestively contain evidence for careful burial with simple grave goods. At Qafzeh the human fossils are variable but certainly include the remains of *Homo sapiens* individuals; and at Skhul most authorities are prepared to accept the fossils within our species despite their retention of a few archaic characteristics (though some of us have reservations). Based on Qafzeh, though, we have incontrovertible evidence in the Levant for humans of modern or effectively modern anatomy at not far short of the 100-kyr mark.

I've already hinted that anatomically modern humans in some way coexisted with Neanderthals in Israel from at least the time of Qafzeh and Skhul to about forty thousand years ago; and it may be significant that the first appearance of the Upper Paleolithic in that region occurred only shortly before the local Neanderthals finally disappeared, suggesting that it was the adoption of modern behavior patterns that finally gave *Homo sapiens* the edge. There's little evidence for early symbolic production, however. A sinuously engraved plaque, clearly a deliberate product, has recently been described from the 50-kyr-old site of Quneitra, on the Golan Heights. At that time the tool kits of Neanderthals and moderns were more or less indistinguishable, and there is no indication whether Quneitra preserves the traces of Neanderthals or of early moderns: it might have been either. But even with the coming of Upper Paleolithic technology, as far as we know uniquely the work of modern people, there does not immediately occur in the Levant any evidence of the symbolic activity so characteristic of early moderns in Europe (whose stone technology was also distinctive). This may be something of a puzzle, but it is clear that there is a wide variety in cultural expression among living peoples; and although symbolic activity is typical of all, it does not always express itself in ways that we might expect to detect in the archaeological record. The early Europeans who saw the Neanderthals off simply happened to possess a cultural repertoire that is bountifully reflected in the amazing record of themselves that they left behind: although, significantly, not equally at every site. Yet it's undeniable that the record we have now suggests that modern behavioral patterns originated much later than the first intimations of anatomical modernity.

Exactly what happened as (behaviorally and anatomically) modern humans invaded Europe is not clear; we don't even know by what route(s) they entered the subcontinent. Until not long ago, things seemed relatively simple: the earliest dates for Upper Paleolithic sites in Europe (invariably associated, where fossils are preserved, with people of modern anatomy) come from the east; in Bulgaria, in particular, the Upper Paleolithic site of Bacho Kiro is dated to over 40 kyr ago. The modern invaders, it appeared, had swept into Europe from somewhere to the east, beginning at around that time. By about 27 kyr ago they had ousted the Neanderthals from their last redoubt on the Iberian Peninsula in the far west. Now, however, we have dates of about 40 kyr for Upper Paleolithic sites in Iberia itself: earlier than for any others west of Bulgaria. So did the early moderns move into Europe at around this time in more than one wave, from different entry points? Did some initial attempts at occupation fail? What does the alternation of the Aurignacian with the Châtelperronian at certain sites in France imply? As new dates come in, the picture will with luck be clarified; meanwhile, it seems that the takeover of Europe by modern people was a complex and, in historical terms, long, drawn-out process, the eviction of the Neanderthals taking over a dozen millennia.

Some paleoanthropologists have argued that the process was not in fact one of eviction but one of absorption: that the incoming moderns interbred with the Neanderthals, whose genes were "swamped" by those of the invaders. Alas, this peaceful scenario finds little support in the archaeological record (which locally shows a pattern of short-term takeover); and the magnitude of the physical differences between Neanderthals and moderns makes it highly unlikely that the two were capable of successful interbreeding. We may admire the people of the European Upper Paleolithic for their artistic achievements; but they, like us, must have had their darker side, and encounters between

Cro-Magnons and Neanderthals cannot always have been happy ones.

All of this, of course, begs the question of the *origin* of modern human behaviors. As we've seen, we find dramatic evidence for art, music, and symbol very early on in the Upper Paleolithic record in Europe, well over 30 kyr ago. We don't know over how long a period modern sensibilities and proclivities were originally acquired; but it's abundantly clear that they were in full flower by that early date. Thus, nobody disputes that the delicate plaques from the Abri Blanchard embody some form of symbolic notation, even though it's not certain that they were actually lunar calendars, as has been suggested. And symbolism lies at the very heart of what it means to be human, as I'll emphasize in the next chapter. For if there is one single thing that distinguishes humans from all other life-forms, living or extinct, it is the capacity for symbolic thought: the ability to generate complex mental symbols and to manipulate them into new combinations. This is the very foundation of imagination and creativity: of the unique ability of humans to create a world in the mind and to re-create it in the real world outside themselves. Other species may exploit the outside world with great efficiency, as we saw in the case of the chimpanzees; but they still remain in essence passive subjects and observers of that world. Even the Neanderthals, remarkable as they may have been, were in all likelihood hardly more liberated from this condition.

It is in the art of the Cro-Magnons that we find the singularly human capacity of these people most dramatically displayed. For, as we've seen, this art was far more than a simple mechanical rendition of the environment that surrounded these early people. It was a complex re-creation of that outside world, rendered with exquisite observation and control. The mythical context(s) of that re-creation we will never know for certain; but what is evident is that even the Cro-Magnons' superb renditions of the animals with which they shared the landscape had symbolic significance to them that surpassed simple zoological identity. We readily recognize the abstract signs with which the animal friezes of Lascaux are peppered as "symbols" (for what else can they be?); but the animal images themselves were clearly more to the Cro-Magnons who painted them than simple representations. They, too, were invested with overtones of the complex world of the mind.

Quite apart from the more ethereal aspects of creativity and imagination, the Cro-Magnons present us with the best early evidence for the complex social organization that is so typical of modern humans. For example, not only do elaborate burials with grave goods such as those from Sungir strongly imply belief in an afterlife, but they also present us with clear evidence of social stratification and the differentiation of social and economic rôles that is so typical of modern human societies. It is inconceivable that all Cro-Magnons could have been so sumptuously buried; and indeed we know that it was not so. Some Cro-Magnons lived humbler existences than others; and while it is likely that some Neanderthals had greater social importance than others, and even that males and females played distinct economic (as well as reproductive) roles, the probability is still that Neanderthal society was radically distinct from anything we are familiar with today. Most Neanderthal sites are small (or if large, consist of numerous small successive living spaces, implying quite severely limited group size at any one time); Cro-Magnon sites embrace a much larger range of sizes, suggesting that, at least seasonally, much larger groups assembled. On the economic level, numerous lines of evidence point toward fundamentally different approaches to exploiting the environment; hafted projectile points are rare at Neanderthal sites, for instance, suggesting that these humans habitually attacked larger prey close-up (an extremely hazardous business, reflected in numerous injuries, still seen on their bones, that are said to resemble those typically sustained by rodeo riders), while Cro-Magnon hurled projectiles from a safer distance.

Interestingly, while it remains unquestioned that some of the earliest Cro-Magnon art was some of the greatest, over time we can readily discern evidence in the Upper Paleolithic record for improvement (as opposed to simple change) on the technological level. The earliest Aurignacian tool kits, for instance, were relatively crude, and the most sophisticated fishing technologies were adopted comparatively late in the Upper Paleolithic. Perhaps this is to be expected, however, for technology is something of an absolute, whose efficiency can be measured quite accurately. Humans routinely discard inefficient ways of doing things in favor of newer and more efficacious developments, and regular technological innovation over the course of the Upper Paleolithic is only the first example we have of a tendency that is still gathering impetus today. In contrast, art expresses a fundamental yearning of the soul in a way whose effectiveness is not directly measurable. Art subserves the same essential function irrespective of how good it is; its excellence is entirely independent of technology; and it changes with fashion in unpredictable rather than linear ways. It's because of this that extraordinary objects such as the Vogelherd horse announce the early arrival of a fully modern human sensibility in a way in which an Aurignacian blade never could; and this immensely ancient and exquisite object will bear eloquent witness to this arrival as long as there are humans on Earth to appreciate it.

Still, although the remarkable record bequeathed to us by the Cro-Magnons is qualitatively the best as well as the most abundant evidence we have of early human creativity, it is far from the oldest. In Africa and in Australia, particularly, we have intimations that modern human behavior patterns were becoming established considerably earlier. Blade tools, for example, show up in Africa over (probably very considerably over) 80 kyr ago; and while one swallow doesn't make a summer, it may also be significant that hafted projectile points are routinely found at African sites in the 100-kyr-plus time range. Recently, barbed bone points have been (somewhat controversially) dated from 60 to 80 kyr ago at a site in Zaire; and there are also hints of bone working at sites in southern Africa of comparable age. What's more, ancient evidence has been found in Africa for flint mining and the long-distance transport of raw materials: behaviors not found in Europe before the Upper Paleolithic. Local differentiation of stoneworking technologies, associated in Europe with the Cro-Magnons, also shows up in Africa at an early date.

All of these things are, of course, related to technological, rather than overtly symbolic, behaviors; but as I've said, the latter are not always of the kind that might be expected to show up in the archaeological record, and, taken overall, the hints of early modern behavioral patterns in Africa are highly suggestive. Direct evidence for symbolic behaviors on that continent is sparser than that for technological innovations, but is not entirely absent: incised fragments of ostrich eggshell are known from sites that may be almost 100 kyr old, and ostrich eggshell beads are known from several widely scattered sites in the 60-to-50-kyr range. Notched bone fragments that may just possibly be notational have been found at sites predating 40 kyr. Thus, although the people of the African Middle Stone Age (roughly equivalent temporally and technologically to the Mousterian) evidently did not have the fondness for symbolic output that the Cro-Magnons had, they nonetheless showed intriguing hints of behavioral patterns more "modern" than those of the Neanderthals.

Australia, too, offers inklings of quite early modern human behaviors. Simply in terms of occupation, Australasia already hosted humans at least as far back as about 60 kyr ago, or maybe earlier yet. This is of more than parochial interest because to get to Australia, early humans had to perform the formidable feat of navigating across at least sixty miles of open ocean. What's more, there are indirect indications from Australia of artistic activity as long ago as 50 kyr or more, and a recent report dates some curious circular rock engravings to much earlier than that, though the dating remains highly controversial. In sum, although outside Europe all we have at present are vague hints such as this of symbolic behaviors, it's clear that at present our perceptions are heavily biased by the richness of the relatively late European record. Evidently, there was a lot more going on in the origin and spread of modern behaviors than has yet met the eye. We'll return to this point in the final pages of this book.

### *Retrospect*

What, then, have we learned of our origins and evolution? Most importantly, it is evident that hominid evolutionary history has not been one of continuous improvement. What's more, *Homo sapiens*, for all of its remarkable capacities, is but one of many twigs on a great evolutionary bush rather than the occupant of a pinnacle that all other species have failed to climb. Since it has not been my concern to dwell here on systematic pattern in human evolution, my account of the acquisition of our human uniquenesses has inevitably had something of a linear flavor to it. For even if descendant species have coexisted—and competed—with their parent species, just as old technologies and behaviors have persisted beside the new, in the direct line of our descent one species—and cultural innovation—has ultimately succeeded another. Nonetheless, it should never be forgotten that the history of the human family eloquently demonstrates that there are many ways to be a hominid and that our way is but one of them. Hominid history has from the beginning been one of continual experimentation, and *Homo sapiens* is unusual among hominids in being alone in the world today.

The human lineage had its origin well over five million years ago, in an evolutionary event that was not an extrapolation of earlier trends, but was rather the product of unpredictable climate change, as the ancestral forest habitat began to shrink. Our remotest ancestors were neither apes in the modern sense, nor human, although the cognitive capacities of the living apes give us our closest insight into what theirs may have been. These early hominids possessed a suite of physical adaptations that is foreign to any living model, for while retaining a facility for climbing, they had begun to acquire the ability to move around habitually on their hind limbs when on the ground. This was not a "transitional" way of life, except in retrospect. Certainly, the australopiths were neither as talented in the trees as the apes nor as facile on the ground as we are; but their adaptation served them extremely well, as the anatomical stability of the group over millions of years attests.

This remarkable stability emphasizes that human evolution from its very beginnings has been episodic rather than gradual. Our ancestors and their relatives were well adapted to the environments in which they found themselves and were highly successful in them, even as the human lineage continued to experiment by producing new species. But not until the next major climatic crisis, an entirely adventitious event, do we find true innovation, with the introduction of the first stone tools. With savanna encroaching yet farther on the woodland habitat of the australopiths, one population of them came up with a completely new approach to exploiting its environment. Exactly how the invention of stone tools affected the general ways of life of these early hominids is hard to say. For we know nothing directly of how their social existences, for example, were affected by this innovation; and although the record is not clear on this point, by conventional reckoning it seems that for a while, at least, these early toolmakers retained their ancestral locomotor adaptations. This observation introduces another feature that is characteristic of our prehistoric record: physical and technological innovations have not gone hand in hand. Although on the surface this may seem surprising, we have to bear in mind that all innovations, of whatever kind, have to arise *within* species; and that members of the same species do not ordinarily show exceedingly large physical differences from each other.

Cognitively, though, the first toolmakers had moved far beyond the apparent capacities of their ancestors. They showed insight into the properties of stone and how it could be fractured to their advantage; and they possessed powers of anticipation that exceed anything seen in the living world outside our own species. Oddly, perhaps, nothing beyond these powers is clearly evident in the record left by the first hominids whose bodily proportions resembled our own. These ancestors were clearly well adapted to an open savanna environment; no longer were the requirements of upright terrestrial locomotion compromised by the retention of physical features that would have been useful in the trees. Here, finally, was a hominid that was completely at home in open country despite the many dangers that this environment carried for creatures lacking the formidable arsenal and the cunning of recent humans. This transformation was a profound one; and it's worth noting in this connection the findings of current research that wholesale reorganization of developmental patterns may arise from relatively minor changes in certain gene complexes. Still, it's unclear at present how such findings relate, if at all, to this radical physical change.

However this may be, a couple of hundred thousand years then elapsed before hominids of this new kind introduced any detectable cultural innovation. This involved a whole new approach to the business of toolmaking: one that focused not simply on an attribute of the stone tools produced—a cutting edge—but on the shape of the tools themselves. To make a carefully shaped hand ax from a lump of rock not only demanded a sophisticated appreciation of how stone can be fashioned by fracture, but a mental template in the mind of the toolmaker that determined the eventual form of the tool. And while at least some of the endocasts bequeathed to us by the earliest toolmakers suggest a limited degree of brain enlargement and asymmetry, with all that these things imply, we only find such features in any significantly advanced form in the brains of the first hominids with whom we share our bodily proportions.

From this point on brains and tools continued to become more elaborate, but in an episodic and unconnected way. Again, what is absent from the picture is any suggestion of continuous improvement. And while physical and cultural innovations continued to be made, both the spottiness and the inherent limitations of the record make it hard to know what they meant in the lives of those who acquired them. Intimations of the control of fire, for instance, go back a very long way; but it was not until relatively recently that this technology became routinely incorporated into hominid life. Only with the arrival of the large-brained Neanderthals, a mere couple of hundred thousand years ago, do we find suggestions in the record of behaviors—such as burial of the dead—that really ring a bell with us. But even here it seems that, innovations of this kind apart, the Neanderthals were doing little more than to refine and improve upon the lifestyles of their ancestors.

For when all is said and done, the Neanderthals did not do a huge amount to bridge the cognitive gap that so strikingly demarcates the recorded behaviors of their predecessors from those of the fully modern humans who succeeded them. In particular, these extinct hominids left behind no unequivocal evidence of

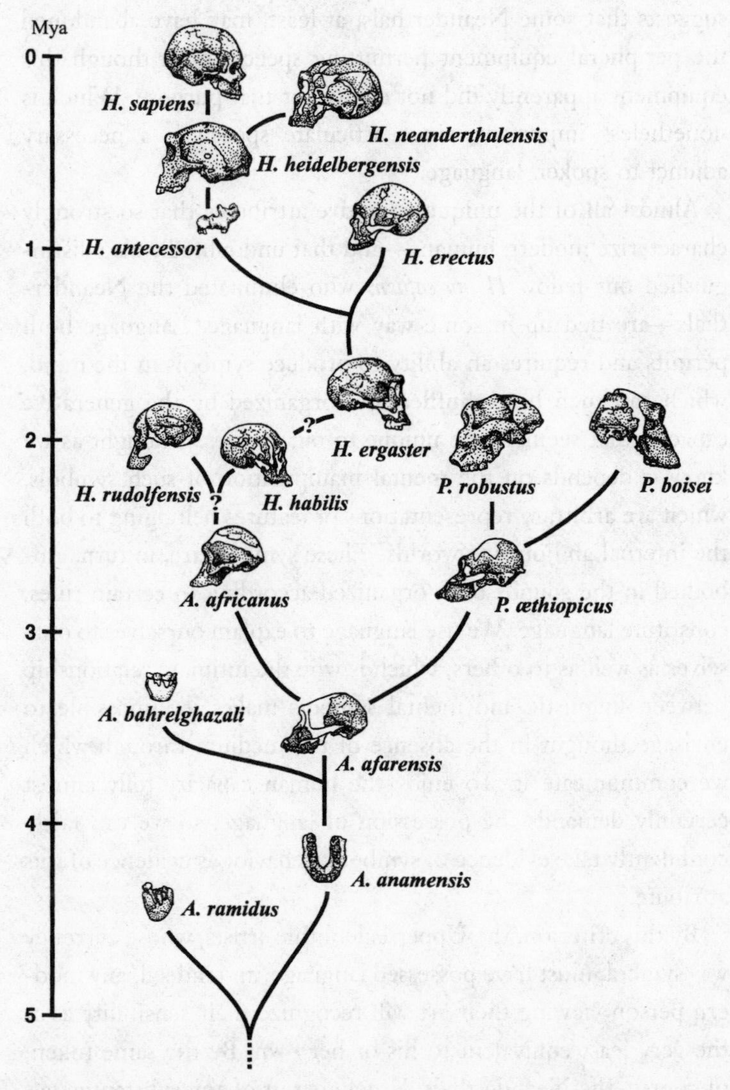


*One possible scheme of relationships among the various known species of the hominid family. The known fossil record is simply the tip of the iceberg, and the true picture is certainly much more complex than this.*

the symbolic behaviors that one might associate with the possession of language. What's more, the anatomical evidence suggests that some Neanderthals, at least, may have abandoned the peripheral equipment permitting speech, even though this equipment apparently did not evolve for that purpose. Which is nonetheless important, since articulate speech is a necessary adjunct to spoken language.

Almost all of the unique cognitive attributes that so strongly characterize modern humans—and that undoubtedly also distinguished our fellow *Homo sapiens* who eliminated the Neanderthals—are tied up in some way with language. Language both permits and requires an ability to produce symbols in the mind, which can then be reshuffled and organized by the generative capacity that seems to be unique to our species. Thought as we know it depends on the mental manipulation of such symbols, which are arbitrary representations of features belonging to both the internal and outside worlds. These symbols are, in turn, embodied in the sounds that, organized according to certain rules, constitute language. We use language to explain ourselves to ourselves as well as to others, which is why the intimate relationship between linguistic and mental symbols makes it impossible to envisage thought in the absence of the medium through which we communicate it. To enjoy the human capacity fully almost certainly demands the possession of language, so we can fairly confidently take evidence of symbolic behavior as evidence of this attribute.

By this criterion, the Upper Paleolithic artists, whose currency was symbol, must have possessed language; and indeed, any modern person viewing their art will recognize their sensibility as at the very least equivalent to his or her own. By the same token, of course, the Neanderthals, largely bereft of any substantial evidence for symbolism despite their manifest other accomplishments, lacked both language and modern sensibility. The

Neanderthals may have been capable of empathy in some way, and what they achieved in difficult environmental circumstances may excite our admiration; but they were not *us* (and why, indeed, should they have been?). From our species' point of view, the crucial cognitive leap—that to complex symbolic reasoning—was quite evidently one that was taken very late in human evolution; and it was achieved by means that are still obscure, though I'll speculate later. This is why if we want to understand what we have become, it is to ourselves that we must look. As it turns out, our evolutionary history is far from irrelevant to this goal. Where it fits in we will explore in the next chapter.

CHAPTER *6*

. . . . . . . . . . . . . . .

# *Being Human*

It may seem odd to follow two chapters expounding five million years of human evolutionary history by saying there's not a great deal we can learn about ourselves by contemplating our evolutionary past that we cannot learn by observing our often bizarre behavior today. Yet there's no doubt in my mind that in one important sense this claim is entirely true. For if the last chapter established anything at all, it showed that with the arrival of behaviorally modern *Homo sapiens*, a totally unprecedented entity had appeared on Earth. For the first time since the adoption of upright walking—or perhaps stone toolmaking—a new kind of hominid was around of which it could not be said that it merely did what its predecessors had done, only a little better, or even just a little differently. *Homo sapiens* is not simply an improved version of its ancestors—it's a new concept, qualitatively distinct from them in highly significant if limited respects. Even if our egotistical species tends to overestimate the size of the qualitative difference between itself and the rest of the organic world, including our closest known relatives, that difference is real. In which case, the extraordinary phenomenon that we are has to be understood in terms of its own unique self—if indeed it can be at all.

Lest it should appear that by claiming this I am trying to put myself and my fellow paleoanthropologists out of business, let me hasten to add that none of what I have just said implies in the least that the study of human evolution has nothing to offer its surviving subject species. It is, after all, hardly without significance that images of extinct humans staring from the covers of *Time* magazine virtually guarantee record sales for the issues thus decorated. A thirst to explain our place in nature and to know where we come from—whether as individuals, as nations, or as a species—is quite evidently one of the most basic aspects of our human psyche; and, just like the origin myths that have doubtless been with us since long before the invention of writing, any new knowledge science can offer about our beginnings helps to satisfy a longing that emanates from the very core of our beings.

Nonetheless, it still remains true that the abilities of today's *Homo sapiens*—and of those extraordinary artists of the Upper Paleolithic—represent a huge leap away from those of our precursors. To put it another way, what has neatly been called the "human capacity" is not simply an extrapolation of the earlier trends in our lineage that the studies of paleoanthropologists are designed to elucidate. It is more akin to an "emergent quality," whereby for chance reasons a new combination of features produces totally unexpected results. The classic example of such a quality is water, whose remarkable characteristics, so essential to life, are entirely unpredicted by those of either hydrogen or oxygen atoms alone. But it may well be that the extraordinary human brain is the best example of all. And it is certainly the emergent nature of our controlling organ and of the abilities that spring from it that make it necessary for us to look to ourselves today—and to the written record, which suggests that in essential respects, human behavior hasn't changed at all since that record began five thousand years ago—to understand what it is that is truly unusual about us. Human beings are so remarkable, and so unprecedented, in all of those features that make us *feel* so different from all of the other myriad and wonderful products of evolution, that only introspection will suffice to tell us who we are. And such self-examination has to start with our remarkable consciousness, from which all else flows.

## *The Uses of Consciousness*

Human beings have endlessly debated the nature of their consciousness, although with equivocal results. The problem is, of course, that this quality is one of inner experience. It is not derived from the outside world, although it has everything to do with the way we perceive that world. Our consciousness is, if you will, the filter through which we view and interpret the environment around us; but it doesn't exist in that environment (which includes all other individuals of our species), and it therefore provides us with no external point of reference on which we can all agree. We *can* agree that we *have* consciousness, but by its very nature, we can't define it in any universal sense. Nonetheless, this capacity is our most conspicuous human characteristic, and it's impossible to ignore it in any account of ourselves.

Where does our consciousness come from? Are our minds distinct from our bodies, or does the one emerge from the other? Most acutely posed by René Descartes well over three centuries ago, this question is still the center of vigorous debate. The introduction of evolutionary thought hardly did anything to resolve it; Charles Darwin was firmly of the opinion that brain evolution through natural selection was the unambiguous explanation of human consciousness, while Alfred Russel Wallace, an energetic proponent of adaptation through natural selection in all other matters, was simply unable to see how this process could have brought into existence the extraordinary awareness of human beings. Today, although the debate shuttles between essentially the same two poles, it seems to me that both of these geniuses were right. Consciousness is the product of our brain, which is in turn the product of our evolution. But the properties of the human brain are emergent, the result of a chance coincidence of acquisitions (based, of course, on the unique product of a long evolutionary history) that could have been favored by natural selection only *after* they had arisen. The mechanisms that lie behind these emergent properties remain among the most important unanswered questions of science, although many lines of investigation are energetically being pursued by neurobiologists, psychologists, philosophers, and others.

One currently popular approach to the problem of understanding consciousness has been to view our brains as machines. For in one limited sense they must indeed be machines, at least to the extent that there is no reason to regard consciousness as anything other than the product of processes that take place within the physical brain. And it is, of course, the inherited mechanical similarities of our brains to each other that permit human beings to assume their consciousnesses are more or less the same and to deal with each other on that basis. The notion that all normal human beings have consciousnesses that are comparable to each other certainly works well enough in practice.

So far so good; but despite the recent blossoming of techniques that have allowed us to gain unprecedented insight into the functioning and dysfunctioning of human brains, we still don't know what the underlying processes are that lead to our possession of what we call consciousness, though hypotheses abound. In theory, an artificial "brain" could be constructed to reproduce the functioning of the human brain, up to and including the experience of consciousness; but we really have no idea how this might be done in practice. Attempts to simulate human responses using computers—to create "artificial intelligence"—have mostly been limited to the development of algorithms that mimic the *results* of "intelligent" approaches to particular problems, although recently the development of "learning networks" has broadened this approach. Some of these endeavors have been extremely impressive, but the essential problem remains. For even if we were unable to distinguish the responses of a perfectly "intelligent" computer from those of a conscious human being (and let's remember that humans come in infinite mental varieties), it's still far from evident that we could conclude that the computer itself possessed consciousness of the kind familiar to us. Conversely, otherwise normal human beings are capable of carrying out complex activities while in "unconscious" states resulting from sleep disorders. Such activities have occasionally landed their practitioners in the courts; but legal decisions haven't done much to help us recognize, let alone define, consciousness: some sleepwalkers have been held legally unaccountable for normally felonious acts committed in this equivocal and poorly understood state, while others have been convicted of crimes. This is also true of defendants afflicted with other brain disorders or whose brains had been damaged in some way.

The notion of the brain as machine (even if not an optimally designed one) will inevitably continue to occupy philosophers and neurobiologists; but so far it has done nothing to help us understand how the brain generates the quality we call consciousness. Clearly, although it may be useful in some respects to study the brain as if it were a machine, it is not a machine in any ordinary sense. For the moment, then, those interested in human consciousness as a product of our evolution will perhaps do better to ask a different question. If we don't know exactly what consciousness *is*, can we usefully ask what it is *for*, or at least, what it allows? I have borrowed the title of this section—"The Uses of Consciousness"—from the psychologist Nicholas Humphrey, who has posed just this question in a particularly elegant way. Humphrey is a strict Darwinian who believes that the human brain has "evolved because, and only because, it is serving some kind of useful biological function." And while it should be clear by now that my view of the evolutionary process is different from his (the remarkable human brain came about and was then favored), Humphrey's inquiry goes straight to the heart of the key problem that confronts everyone interested in the origin of modern human consciousness. I would only rephrase the question slightly: What was it about the mental processes of the species possessing this capacity that allowed it to triumph over its relatives that didn't?

Humphrey's answer to this question is related to his metaphor for what he regards as the principal feature of human consciousness. In his view, consciousness is provided by an "inner eye": a unique property of the mind, based on who-knows-what structural or physiochemical attributes, that allows the brain to observe itself at work. This ability—which we may assume is temporarily suspended in true sleepwalkers and is absent in individuals with certain kinds of brain damage—permits what Humphrey calls "self-reflexive insight." I must admit that I find Humphrey's metaphor persuasive, even if it does not embrace every aspect that common sense might attribute to consciousness. I follow him, too, in concluding that "the depth, complexity, and biological importance" of human interpersonal relationships, which "far exceed those of any other animal," would be impossible without the capacity for self-reflexive insight. True, we have seen that the double bluff of chimpanzees may well indicate a certain rudimentary ability to read another's mind; but there's no doubt that Humphrey is right in saying that in the case of human beings, "without the ability to understand, predict and manipulate the behavior of other members of his own species, a person could hardly survive from day to day." To an altogether unprecedented degree, insight into the motives of others—impossible without some degree of insight into one's own—is an essential ingredient of human social behavior.

To a strict Darwinian, this observation is enough. Human consciousness has simply—and virtually inevitably—resulted from the reproductive advantage conferred by natural selection, generation after generation, on individuals with ever greater powers of self-reflexive insight. But we've already seen that the evolutionary process is actually a lot more complex than this, and among highly social animals such as the primates, it's hard to see why the same thing has not happened in all lineages if tiny behavioral advantages are inexorably amplified over time in this way. What's more, there is no better example than the history of the vertebrate brain to demonstrate that evolutionary change has not consisted merely of gradual improvement over the ages. Brain evolution has not proceeded by the simple addition of a few more connections here and there, finally adding up, over the aeons, to a large and magnificently burnished machine. Opportunistic evolution has conscripted old parts of the brain to new functions in a rather untidy fashion, and new structures have been added and old ones enlarged in a rather haphazard way.

To see clearly in this matter, we have to recognize that many levels of natural process were involved in getting us to where we are today. First, the modern human brain arose, within a local population of early humans and through developmental alterations we don't yet understand, from a precursor form that possessed a variety of necessary exaptations. Next, natural selection worked within that local population to establish this new variant as the norm. Then speciation intervened to establish the historic individuality of this new entity. And, finally, the new species competed successfully with its relatives, in a process that eventually left *Homo sapiens* as the only hominid on the scene—perhaps for the first time since not long after the ancestral hominid originally appeared. Viewed this way, fully human consciousness is simply one more effect of the routine and random emergence and fixation of innovations that occurs in the evolution of all lineages.

What do we conclude from this? First, we have to recognize that once the modern brain and consciousness had been acquired, a whole package of cognitive potentials was available to be exploited, if not necessarily immediately. I am fully in accord with Humphrey that the complex social relationships made possible by the "inner eye" express a fundamental component of the modern human consciousness. I am also happy to surmise that such relationships were not characteristic—at least in fully developed form—of our immediate precursor. But I am dubious that the "inner eye" (at least as it relates to interindividual relationships) provides a unique key to the success of our species; for, rather than being the cause of our consciousness, it is more plausibly simply one result of our capacity for symbolic abstraction. Moreover, while such a quality might well promote the success of individuals *within* species by making them better readers of the minds of their companions, it's not necessarily obvious how by itself it would provide a competitive advantage *among* species. And, quite simply, there is clearly more to our flexible and diverse cognitive capacities than simply the ability to interact with each other in subtle and complicated ways, remarkable though these are. Indeed, if we have to identify any single characteristic that sets us apart, one of the things that is truly extraordinary about human beings is their finely honed perception of the world beyond their social milieu.

Painstaking and beautifully designed research by Dorothy

Cheney and Robert Seyfarth, for example, has shown an altogether unexpected subtlety and complexity in the social relationships of vervet monkeys; yet these primates show considerably less insight in dealing with the environment around them. For example, they respond poorly to cues about predators in the habitat, and in finding food they fail to show the cooperation and ingenuity that they display so conspicuously in social contexts. Chimpanzees do rather better in this regard, but we can nonetheless identify with them much more closely as social creatures than as planners and careful readers and exploiters of the environment.

In our predominantly urban and universally high-tech environment, we are besieged by social stimuli, and our interactions with the nonsocial world are largely dominated by our relationships with our own technology. I am writing this book using a word processor; I shall travel home by subway; and, once surrounded by familiar artifacts, I may well relax by listening to music recorded on a compact disc. Yet even now, we are not far (in time, at least) from our roots as incomparable readers of the external environment: the part of our total environment that we have not ourselves created. Recently I met a middle-aged French bus driver who fascinated me with accounts of his youth in the countryside of the Corrèze, not that long ago. His best friends were older men, poachers. These men could peer through a rustling forest and unerringly identify behind which of many waving tussocks a frightened and invisible rabbit crouched. Better known in our own days of *National Geographic* specials are the Australian Aborigines, who can survive for extended periods in apparently featureless and waterless deserts in which you or I would promptly perish; or the San Bushmen, who can tell from a bent twig or a stone overturned in a streambed what animal had passed, in which direction, how long ago, how fast it was traveling, and whether it was wounded. It is a true uniqueness of human beings to be able to read subtle signs of this sort in the world outside themselves and to exploit those signs to their advantage. If the acquisition of the modern human capacity conferred any single preponderant benefit on *Homo sapiens* as a competitive element in nature, my guess is that it was surely this. How this capacity was released is another question, of course, to which I'll return at the end of this chapter.

## *The Human Noncondition*

From capacity comes condition; and over millennia now, philosophers and theologians have made something of an industry of debating the human condition. Even if inevitable, it is rather ironic that the very species that apparently so much enjoys agonizing over its own condition is, in fact, the only species that doesn't have one—or at any rate, whose condition, if any, is most difficult to define. Whatever condition is, it is surely a lot easier to specify it in the case of an amoeba, or a lizard, or a shrew, or even a chimpanzee, than it is in our own. Chimpanzees are complex creatures, of that there is no doubt; but the range of possibilities open to a chimpanzee is hugely more limited than the options confronting any individual *Homo sapiens.* Of course, we are also confronted by many of the same fundamental realities that chimpanzees face, none of which can be ignored in any attempt to define our condition. But there's obviously nothing specifically human in the problems we share with chimpanzees; rather, it is the way in which we are conscious of and deal with the realities of life that separates us. And, much as we prize our powers of reasoning, it is not the unbridled exercise of rationality that distinguishes us in these pursuits; after all, the history of mankind is littered with the slaughtered carcasses of golden geese, as its future will continue to be. No, it's a lot more complicated than that.

To humans, life is not simply an economic business. It is overlain with all kinds of social and symbolic complexities, and we have discovered literally thousands of ways both to create and to deal with these. Each society has invented its own ways of coping with economic and social needs, and with the knowledge of individual mortality. What's more, appalled though members of one society may be by ways of doing business in another, no society is intrinsically better or worse than others in any universal moral sense. We can derive no concepts of morality (a social construct) or of "natural law" (an intellectual construct) from the contemplation of nature—or of any other material entity that lies outside our corporate selves; for nature is simply indifferent to individual suffering or success, and to call such indifference amoral would be simply to anthropomorphize. Thus, for example, while I passionately believe that any moral system is deeply flawed that does not start from the proposition that the individual is the unit of suffering, I have to recognize that I can find no justification for this notion in anything that lies beyond the human milieu (or perhaps just my own perception of it). Moral codes and behavioral norms are basic necessities for such individually complex and mysteriously motivated, yet highly social beings as we are; and in deriving them we must necessarily look to ourselves. The results of such introspection vary greatly, though, depending on where exactly we look; and we certainly cannot describe standard practice in any given society as a violation of the human condition. The upshot is that, as a species, *Homo sapiens* presents a bewildering variety that is next to impossible to boil down to a neat account of anything we could describe as *the* human condition, unless we factor in such elements as language and symbolic thought, critical factors that we'll discuss later, and that in any event probably embody cause rather than effect. We are both individual and social creatures; and on the social level, the most basic component of our condition—and what principally necessitates systems of morality—is that we have to live with each other, complex and unpredictable beings all.

But even this is begging the question, for who are we? *Homo sapiens*, as has been recognized since time immemorial, is a bundle of paradoxes, individually as well as corporately. Let's leave aside societies for the moment, since each culture has simply made its own selection from the vast range of values and behaviors available to *Homo sapiens* as a whole. How about individual human behaviors? Well, these can be described by any pair of antitheses you care to think of: generous, selfish; gullible, shrewd; aggressive, retiring; smart, dumb; kind, cruel; shy, assertive; and on and on. Even more significantly, such contradictions can—indeed, to some extent almost invariably do—exist within the same body. Consider this excerpt from the dust jacket of a biography: "loving husband and father, dissipated whore-chaser; conscientious lawyer, drunken buffoon; writer of tedious doggerel, and one of the finest biographies in the English language; what is one to make of a man such as this?" Well, what is one to make of anybody? As they say in the north of England, "there's nowt so queer as folks," which is probably the closest we'll ever get to an accurate description of the individual human condition. The man just described was James Boswell, Johnson's biographer; and although Boswell may have been a modestly dramatic example of the human propensity for inconsistency, he shared that quality with virtually everyone. How many times could such a description be repeated? Evangelical ministers rail in public against adultery and social permissiveness, while living private lives of total moral squalor; politicians preach leaner government while reaching into the pork barrel; even the monstrous Adolf Hitler is said to have been kind to children and dogs. How many human actions do we find ourselves compelled to describe as "inhumane" or "inhuman"? What we are as humans, and what we like nowadays to think we are, evidently are two very different things—as any visitor to the dungeons of a medieval castle or observer of recent events in Rwanda will attest.

Looking for any specific universals in seeking the human condition is thus plainly a fruitless endeavor—although "cognitive dissonance" perhaps comes close. We each have a menu of possibilities—a menu whose full extent it is probably impossible for any one individual to imagine—from which to choose our behaviors. Not, of course, that such choice is necessarily conscious, although we are certainly able to modify our outward behaviors as the situation demands: we are rarely rude to our bosses, whatever we may think of them, because we know it would be unwise. But it is not always as clear-cut and situational as that. Why do some people feel so threatened by the private beliefs and activities of others? Why do so many exhibit such inappropriate senses of entitlement? Why are we so often impelled to caricature sensible ideas by taking them to ludicrous extremes? Why are we so frequently aggressive when we know we are in the wrong? Why Schadenfreude in the absence of any concrete benefit to the observer? Everyone will provide some kind of rationalization for behaviors such as these, but in virtually every case that rationalization will have little if anything to do with the real source of those feelings. In these instances, the "inner eye," so crucial to our cognitive uniqueness, is producing distortions, founded in our innate ability to believe our own lies—which are often concocted after the fact, to suit our convenience. And if we believe our own lies, there's little wonder that we so often believe the lies of others, especially if it is advantageous for us to do so or if they embody notions, however implausible, that we *want* to believe. Such self-deception provides a sort of social feedback that leads to bodies of myth, which in turn amplify the resulting behaviors—and on which, even more destructively, so many established beliefs and behavioral conventions are based (in our own society think among other things of witchcraft, which was actually practiced by nobody; if anybody does it today, it is a leisure activity based purely on myth).

The most basic reason why we are reluctant to acknowledge the contradictions of human nature is, of course, that we have to live together; and if we are to do so, we have to subscribe (at least publicly) to a set of common values and behavioral norms (which may sometimes be contrary to our private beliefs or impulses). But on the social level, defining the permissible range of those values and norms has turned out to be an enormous problem, exacerbated by the traits I've just mentioned. In our society, and many others, behavioral values are founded in a moral code that was established initially in a religious context (though cause and effect here are intricately intertwined). And because every human society possesses religion of some sort, complete with origin myths that purportedly explain the relationship of humans to the world around them, religion cannot be discounted from any discussion of specifically human behaviors. What's more, religion in some sense is one of the earliest special proclivities that we are able to detect in the archaeological record of modern humans; for even if we do not understand precisely what the artistic productions of the Cro-Magnons represented to the people who made them, it's nonetheless clear that this art reflected a view of these peoples' place in the world and a body of mythology that explained that place. Along with a deep desire to deny the finality of death and a curious reluctance to accept the inevitable limitations of mundane human experience, the provision of such explanation is today, and almost certainly always has been, one of the major functions of religious belief. And it's precisely because the art of the Cro-Magnons so clearly goes well beyond pure representation, to embody a broadly religious symbolism, that we are able to identify so closely with these long-vanished people.

But I raise the issue of religion here mainly because, ironically, it is in our notions of God that we see our own human condition most compactly reflected. In this I inevitably come back to an observation I made earlier. Human beings, despite their unique associative mental abilities, are incapable of thoroughly envisioning entities that lie truly outside their own experience or that cannot be predicted from what they know of the material world. God, surely, is an entity of such a kind. In the days when the founding documents of the world's major religions were written, the universe was a familiar, if sometimes hostile and unpredictable place; for as far as almost everyone was concerned, it was limited not just to Earth itself but to the local region. The stars may have moved in the sky, they may have had mystical properties attributed to them, and they may have allowed the calibration of the seasons; but apart from this limited role of the heavens, to all intents and purposes, human notions of where our kind fits into nature were determined solely by the local surroundings and experience of each human group: a comfortingly familiar context. In the week in which I write, though, the Hubble Telescope has expanded the universe by an estimated forty *billion* galaxies—and ten billion were already known. In this unimaginably vast expanse, Earth and its inhabitants are the merest, most microscopic speck; and our concepts of God surely need to expand commensurately.

Yet, even at the end of the twentieth century, our notions of the Creator remain, as they always have been, resolutely anthropomorphic. Our gods are by turns savage, or loving, or jealous, or merciful, or perhaps all at once: a reflection, indeed, of our own contradictory natures. According to one version of the story, we were created in God's image; and if He is indeed of this kind, we certainly were. More likely, though, we are construing God in our own image simply because, no matter how much we may pride ourselves on our capacities of abstract thought, we are unable to do otherwise. To describe God in terms of our own understandings and attributes, understandable though it is, seems somehow demeaning to Him; yet what choice do we have if we wish to believe in a God with whom we can have a personal relationship—indeed, any humanly comprehensible relationship at all? God like ourselves—yet noncorporeal, able to do things that we can't even conceive of doing, and in command of a universe whose immensity we cannot comprehend? There seems to be a typically human paradox here. Perhaps we will one day be able at least to admit of a God possessing sufficient majesty and expansiveness to transcend the limits of our own imaginations and experience. But meanwhile, although we need not literally heed Wittgenstein's admonition not to speak whereof we do not know—after all, how much fun would that be?—we might do well to look upon the inadequacy of our concepts of God as the truest mirror of those limitations that define our condition.

## *Our Evolutionary Heritage*

Whatever human nature is, a fundamental aspect of it seems to be a thirst to know more about ourselves, and in particular to explain our often bizarre behaviors. One recently fashionable and widely publicized approach to doing this is based on George Williams's ultra-Darwinist view of the evolutionary process, refined by reference to the notions of kin selection and the selfish gene that I have already discussed. The resulting discipline of evolutionary psychology, essentially sociobiology as applied to our species, purports to explain vast amounts of human behavior by reference to our genetic heritage, implying that, in a very concrete way, we are prisoners of our evolutionary past. In a general sense, our past is, of course, always with us, as the sufferings of those with slipped disks—the inevitable result of adapting a quadrupedal spine to upright posture—attest. But the emergent nature of the human capacity renders more than somewhat suspect the notion that our behaviors are programmed in any detail by our genetic heritage. Nonetheless, it's worthwhile to pause here to look briefly at evolutionary psychology. Partly this is because notions of this kind are highly attractive to a species that loves simple, sweeping explanations, with the result that this approach to understanding ourselves has received a lot of fawning publicity. And partly, of course, it's because it is obvious that our behaviors cannot be totally unrelated to our genetic makeup. Here it's important to bear in mind that, while evidence is rapidly accumulating that many of our individual personality traits (as opposed to specific behaviors) are closely conditioned by inheritance, it's tricky to generalize from this, as evolutionary psychologists do, to our species as a whole.

To an evolutionary psychologist, the study of human behaviors—universal human behaviors, not simply their manifold social manifestations—starts with a search for "mental adaptations" that have resulted from the action of natural selection in our species' past. When certain social stimuli produce a particular behavioral result (or even merely tend to or are perceived to), the question always posed is this: What evolutionary advantage did this response provide while we were evolving? (which means, to evolutionary psychologists, in a hunting-gathering context). Here we have an extremely reductionist notion of the evolutionary process, in which virtually every species attribute—in this case, myriad human behaviors—has been directly under the control of natural selection over many thousands of generations (until such selection mysteriously went into abeyance with the Agricultural Revolution). While, as we've seen, this sociobiological scenario evades the multifarious complexity of the evolutionary process, it is undeniably attractive to the human mind in its elegant simplicity; and, through its efforts to reject the horrors of social Darwinism and its various unattractive derivatives such as eugenics, evolutionary psychology has performed the useful service of making it respectable once again to discuss human behavior in biological terms. In fairness, it should also be pointed out that most evolutionary psychologists are well aware of the role of individual experience and social constraints—or their lack—in shaping behavior.

In their search for mental adaptations, evolutionary psychologists are fond of referring to the "ancestral environment" in which our own behavioral adaptations evolved. How or where we live today is irrelevant, they say, because industrial or even agricultural lifestyles are such recent phenomena. For keys to our mental adaptations, we must look to our hunting-gathering past. Well, as I've intimated, there is a whole host of problems here, of which let's take only two. First, even if our knowledge of the evolutionary process teaches us that our species has to have originated in a single local area, and thus in a fairly specific environment, our *success* has had to do with our remarkable ability to move out from that environment and displace our competitors from all other human habitats. Evidently what we innately are has little to do with specific adaptation to any particular environment, whether social, geographical, technological, or ecological.

The second consideration, possibly the more important one here, concerns lifestyle and its relation to the selfish gene and associated notions that emphasize the importance in evolution of individual gene transmission. Our early ancestors were in some sense hunters and gatherers, of that there's no doubt. And while hunting and gathering lifeways can be quite diverse, all have this in common: that for people consistently on the move, and exploiting only that which the environment has to offer, large numbers of slow-developing children are no boon. Females are severely handicapped by having to care for more than two offspring at a time, with no beasts of burden to assist in their transport. Indeed, among the few remaining modern hunter-gatherers, women often go to considerable lengths to limit pregnancy. San women who roam the Namib Desert, for example, breast-feed their infants until the latter reach age four or more, thereby keeping their prolactin levels up and inhibiting ovulation for all that time. Their genes hardly seem to be screaming out for replication; and economic considerations, as virtually always, lie to the fore. For hunters and gatherers, then, it's fertility, not its lack, that is the enemy. Individual San women show no sign, conscious or unconscious, of wishing to maximize their output of progeny. And that, folks, is the ancestral environment.

This hardly comes as a surprise, of course; for it has long been realized that, in common with many other organisms, humans have adopted a reproductive strategy that involves high parental investment in a relatively small number of offspring. The goal has never been to maximize the number of offspring per se, but rather to assure the success (economic in the first place) of those offspring produced. The unprecedented demand for labor that followed the adoption of a settled existence and agriculture may have perturbed this pattern a little; but the basic principle persists.

There is a host of other problems associated with seeing organisms simply as vehicles for their genes, rather as in the view that a chicken is merely an egg's way of making another egg. The difficulties are greatest when this outlook is generalized to the structure of complex societies of the primate type and to the behavior of individuals within them. Many such difficulties result from a failure to appreciate that societies have economic as well as reproductive functions and that individuals are likewise economic as well as reproductive entities. Indeed, the amount of time and energy each individual invests in economic pursuits over his or her lifetime vastly outweighs what is spent on reproductive activity. Most of the assumed regularities in human behavior to which evolutionary psychologists point are at least as plausibly due to rational economic decisions (amplified by social norms that also result from concerns we may broadly characterize as economic) as they are to inherited behaviors. When behaviors are observed that do not conform to sociobiological prediction, evolutionary psychologists, who evidently like to eat their cake as well as to have it, are wont to say somewhat paradoxically that in these cases natural selection has gone for "flexibility" instead of specificity in behavior. Once again, though, such behaviors are much better accounted for by the complex nature of the economic decisions (broadly defined) that each organism has to make, combined with the huge variety of personality types found within every human population. Individual genes cannot and do not directly equal behavior, even if corporately and in combination with other factors they clearly play a role. The genome is not that simple; organisms are not that simple; and the social and economic milieus are not that simple, either.

Given their emphasis on gene transmission, it's not surprising that evolutionary psychologists have concentrated a lot of attention upon human sex-related behaviors. This has led them into a particularly tricky area because among humans sexual activity and reproduction are not related in any simple way. Not that this decoupling (as it were) is a purely human phenomenon: bonobos, for example, regularly use sexual activity as a way of relieving social tensions that develop within their groups (and, interestingly, do not show male dominance, suggesting that, to the extent that hominoid male and female behaviors differ, these differences are not necessarily simple by-products of the hormonal systems that govern reproductive function in each). In both humans and bonobos, most sexual activity is unrelated to reproduction, and among humans the interaction of sex with economics is particularly complex. Let's take one example. It is no secret that males and females have different roles in the reproductive process and often display different attitudes toward it. An evolutionary psychologist, assuming that sex and reproduction are effectively the same thing, would say that these attitudes differ because females are a scarce reproductive resource for which males compete. Males are potentially able to impregnate lots of females, whereas females are strictly limited in the numbers of offspring they can have. Hence women's alleged tendency to focus on bonding with one male who can contribute (as munificently as possible) to the rearing of those few offspring, while males are notorious philanderers. The alleged goal in each case is to maximize the efficiency of gene transmission.

"Infidelity: It May Be in Our Genes" trumpeted the cover of a national newsmagazine recently, reflecting how surprisingly easy it has been for evolutionary psychologists to sell this notion that in the all-important effort to ensure their genes' success, men attempt to sire the largest number of children while females are programmed to reserve mating for those males who will invest most in those offspring. These conflicting strategies call for a maximum of deceit on all sides, a notion that appears to be particularly attractive to a society where soap operas of various kinds represent the most lucrative of the televisual arts if not the one that most closely mirrors the realities of most people's lives. But a moment's thought is enough to show that gene transmission pure and simple cannot be the whole story. Impregnating as many females as possible is of little long-term use to males obsessed with the immortality of their genes if their time- and energy-consuming philandering activities detract from the effort they might otherwise have made to ensure the survival and success of their offspring. This trade-off is usually unavoidable, and it is hard to see how "selfish" genes might benefit by it. What's more, even in such relatively simple organisms as fruit flies, it now turns out that literally dozens of unassociated genes are involved in courtship behavior, which in itself comprises only a fraction of total reproductive activity. Behavior, especially in humans and other primates, is no simple matter; and, however seductive the notion may appear that specific genes exist "for" any of those behaviors beloved of evolutionary psychologists, it is woefully wrong.

What would an economist make of all this? He or she would probably note that sex is in itself pleasurable for both sides (which is a species-wide and presumably highly selected–for phenomenon, for in its absence the species would go extinct) and that this in itself explains the male desire for lots of sex, most of which has nothing to do with reproduction—as witnessed, for example, by the venerable but always-booming business of prostitution, an economic venture if there ever was one. Females, however, have to bear the economic consequences of an activity that leads to reproduction and the associated costs of child rearing—something in which males in theory participate only voluntarily. For females, then, the pleasures of unlimited and varied sex may be outweighed by the economic advantages of bonding with a male (provided that the male plays his appointed supportive role, which all too often seems not to be the case, as the current torrent of reports of wife and child abuse appears to suggest). It cannot be insignificant that where the standard economic calculation has been reversed (as when child-support payments are made to unpaired mothers), female bonding behaviors have tended to decline markedly.

Obviously, I don't wish to suggest that males and females do not on average show a variety of gender-specific tendencies; certainly they do, and it even seems likely that male and female brains will turn out to display subtle differences in many aspects of their organization. But by averaging out behaviors, the evolutionary psychological approach vastly oversimplifies the complexities of individual human experience. We may indeed perceive certain overall regularities in human behavior; but usually these are simply the sum of millions of individual decisions (admittedly conditioned by learned values), just as are the unquestioned regularities in the free-market system that economists perceive. In reality, people individually are much more mysterious, complexly motivated beings than evolutionary psychologists (or, to be fair, most economists) suggest. Our existences are rarely blindly devoted to the struggle to maximize our reproductive success, as witness the 10 percent of married couples in our society who voluntarily choose childlessness or homosexuals who consciously reject the option of passing along their genes (although, in view of the increasing frequency of homosexual adoption, it appears that some maintain a strong desire to nurture). Life is, indeed, mostly a matter of keeping alive and as comfortable as possible in a staggeringly complex social environment. This is a supremely economic business. Sex often intrudes, of course, but most often in contexts of status and power, of cementing social bonds, or purely of recreation. It is practiced much more rarely in the specific context of reproduction.

Of course, none of this implies that genes are irrelevant to any aspect of our biology. But while our genes are always lurking within, they only intervene in our behaviors in an indirect way, by programming the development of our brains. These mysterious organs are, of course, the result of a very long evolutionary history; and it is in this sense that the past is still very much with us. If we are to understand the complexities of our behavior, it is to our brains, not directly to our genes, that we have to look.

## *Brains and Behavior*

We are psychologically so complex—I almost wrote "screwed up"—at least partly because of the way in which our brains were built up over the ages, structure on structure. I have already noted that the old notion of an inherent conflict between older and newer brain structures and functions now seems oversimplified; but it is nonetheless self-evident that it is in our controlling organ, itself the product of a long evolutionary history, that we must search for the keys to the contradictions we all exhibit, every day of our lives. At one level our faculties tell us that death is final; but at another we reject the notion and grasp at the most improbable alternatives. We are simultaneously nostalgics and neophiliacs. We commit hate crimes in the name of a loving God. We invent lies and we believe them. We want autonomy in our own lives, but we want to interfere in the lives of others. We have marvelous rational powers, but we follow the dictates of reason erratically at best. Examples of our illogicality are endless. Why?

One possible solution is that such contradictions stem from the complex interaction that goes on in our brains between our higher cortical areas and those more ancient structures that lie below them. As we've seen, though, there are difficulties with this view; for functionally, these older and younger systems are tightly intertwined, and it is the total structure of our brains that provides us with our less rational proclivities as well as with those abilities that allow us to design supercomputers and to send spacecraft to Mars. What's more, in behavioral terms as well, it appears that it's highly misleading to make a simple dichotomy between emotional and rational brain functions. For between the two, there exists an intermediate level of neurobehavioral function, with a foot in both camps. This is intuition, which acts in the absence of conscious reasoning and has long been considered by many cognitive psychologists to have its roots in emotional memory. And now a strong suggestion has emerged that intuition also plays a major role in rational decision making.

When individuals make decisions, their brains do not simply run through a standard algorithm that weighs the facts and produces an optimized solution. Instead, the rational weighing of the evidence is supplemented by all kinds of other inputs that may range from generalized unease or confidence to outright fear, or a nagging feeling that one or another decision will be the right one. The resulting conclusion, while arrived at by pondering the facts, may not be one that the individual can rationalize verbally—or if he or she can, it may be just that: a rationalization. In a very ingenious attempt to elucidate this level of mental function, Hannah and Antonio Damasio and colleagues at the University of Iowa College of Medicine recently asked why patients with damage to the ventromedial prefrontal association cortex, the area of brain lying directly above the eyes, could do well on memory and IQ tests, yet would typically make dreadful decisions in normal life—and have a hard time making them. They set up a gambling experiment in which different decks of cards provided different chances of winning: a situation that only became clear to their subjects after considerable experience using the different decks. However, normal "control" participants quite quickly began to favor the winning decks, well before they were able to say why they were doing so; patients with prefrontal lesions, in contrast, continued to take cards from the losing decks even after they had figured out which decks were the winners. All the participants were monitored for skin conductance response, a measure of psychological stress similar to that used in lie detectors. Those with lesions showed no evidence of stress, whatever decks they chose, while, in contrast, controls rapidly showed stress responses when contemplating taking cards from the losing decks, even before they had consciously identified those decks as the losers.

The conclusion from all this was pretty clear: Nonconscious

biases—intuitions—guided the behavior of the controls before they gained conscious knowledge of which decks were winners or losers. The patients with lesions, on the other hand, were just as capable of figuring out which decks were good and which bad but did not possess the intuitive faculty. The inability of the patients to make good life decisions was presumably related to their defective intuition, which in turn stemmed from their prefrontal lesions. The Damasios believe in the light of these results that the ventromedial prefrontal cortex forms part of the system that stores information about previous rewards and punishments: information used to stimulate the intuitions that then enter into normal people's decision-making processes. All of this makes excellent sense, for it's clear (intuitively obvious?) that intuition is an indispensable pillar of human creative achievement. Even science, the epitome of human rational activities, gives us evidence of this. For while science prides itself on the objective testing of hypotheses that have emerged from carefully collected observations, the initial hypotheses themselves are as often as not the product of insight rather than of verbally based ratiocination. Intuition is thus an indispensable mediator of our thought processes. On the other hand, of course, it could be pointed out that while great music always has an intellectual component, its power invariably derives from its ability to touch our feelings directly, bypassing our rational faculties, and our intuitions, too (unless the music itself has extraneous associations). Nevertheless, even here we must be cautious: recent research suggests that early exposure to music (playing more so than listening, though both are effective) enhances the development of mathematical reasoning, apparently stimulating the development of neuronal circuits devoted to higher reasoning. This underlines yet further the intricate relationship between what we call our lower and higher mental faculties. In the aggregate, though, it's clear that human cognition is composed of what we can broadly categorize as emotional, intuitive, and conscious ratiocinative processes. And it is very likely the addition of symbolic reasoning to preexisting functions of emotional response and intuitive insight that marked the final leap to modern human consciousness.

Nevertheless, we did not leave the past behind with our great leap forward. For while the conflicted behaviors that mark the experience of every human individual may not always stem directly from the oft-made intellect/emotion dichotomy, it is quite likely that intuitions, based on emotional reactions to previous experience, often contradict the rational, symbolic evaluation of the available facts. It's notable in this connection that although we are readily able to learn technological skills by absorbing the results of others' experience, in matters of human interaction we are more or less incapable of learning in this way. No matter how much wisdom in the conduct of human affairs has been passed down through the ages, each generation has made the same mistakes over and over again. Combine this propensity with the inherent limitations of human choice, and it's easy to see why history tends to repeat itself on such a short cycle. Readily and permanently as we learn from someone else how to carry out such technological chores as programming the VCR (apparently it can be done), in our interactions with others we seem able to learn only by the most powerful of aversive conditioning. Only by dismal personal experience do we grasp the wisest ways of dealing with each other; and even then we find it hard to generalize our hard-won experience to new situations. Right brain/left brain isn't the major question here, even though it is what's had the publicity; the real issue is the interaction of several diffuse brain modalities, which *together* make the mind as we experience it: a potential source of tension from which, as we'll see, we shall never be freed.

It is here, then, in the intricate structures of our brains, that the assumptions of evolutionary psychology truly break down. What our remote human ancestors did, if not entirely irrelevant to our modern condition, is hardly to be blamed for our behaviors today. Whether these ancestors were hunters or gatherers, or both; whether they lived in large or small groups; whether early human males were promiscuous, or faithful providers—all this has little bearing on what we are (or can be) in the late twentieth century. Our brains are not identical machines that are programmed by the genes to respond in specific ways to specific stimuli; the multifarious cultural differences that exist around the globe are proof enough of that, as are the extraordinary differences among individuals within each society. Rather, we are behaviorally complex beings, exquisitely responsive to experience, among whom internal conflict and compulsive behaviors are often as much the norm as the exception. This behavioral complexity has doubtless been with us since the birth of modern *Homo sapiens;* but ancient lifestyles have little to do with how we live our lives now.

Our rational minds, moreover, are not perfect devices, ever on the alert; even at our most detached, we often make mathematical errors or wrong decisions about how to deal with new aspects of this complex world we have made for ourselves. Human error is an unavoidable reality of human existence; yet—contradiction again—we howled for the Exxon Corporation's blood at the time of the *Exxon Valdez* disaster, a supremely statistical accident—while knowing all along, had we cared to admit it to ourselves, that the true culprits were every one of us, each time we switched on the pool filter or started up our gas guzzler. What's more, even when we know, in principle, how to act in the best interests of ourselves and others, few if any of us are able to put this knowledge consistently or perfectly into practice.

Everyone can act rationally some of the time; but nobody can act utterly rationally all of the time. And, despite the evident downside, we should probably be thankful for that: love and compassion are not, after all, purely rational qualities.

## *Humans and the World*

Human beings are a part of nature. We are the result of the same natural processes that have produced all of the other myriad life-forms in the world. Yet we *feel* different from our fellow organisms, and of course we are—although it can be argued that all other species are, too. But we are truly different in the fundamental sense that we, and we alone, are able to reflect on that difference. And our views on the nature of such difference inevitably impinge upon our notions of our place in the world. At this point, the matter leaves the pure realms of philosophy or theology or biology and takes on an intensely practical significance; for how we perceive our role in the world strongly governs the ways in which we interact with it.

In his recent book *Dominion*, my colleague Niles Eldredge points out that there seems to be a fundamental difference in attitudes toward the outside world between surviving hunting-gathering peoples and those who have adopted an agricultural way of life. By way of illustration, Eldredge quotes the anthropologist Colin Turnbull's observations on the Pygmies of Zaire's Mbuti Forest. When these people leave their homes to forage (or, more correctly, to collect), they sing out to "Mother Forest, Father Forest," an activity which Eldredge takes as emblematic of the Mbuti peoples' integration into the ecosystem around them. The Mbuti are highly skilled exploiters of their environment, but the key to their relationship with it is that they don't seek to modify it in any way. They know this ecosystem intimately and have an astonishingly detailed knowledge of the resources it offers them; but they are part of it, and depend upon it, and understand this dependency. This is reflected not only in their descriptions of the forest itself, but in their accounts of themselves and of where they stand in relation to the nature that surrounds them.

Something similar must have been true of all the many hunting-gathering peoples who have existed on Earth—and until quite recently there were *only* hunting-gathering peoples (although, admittedly, this term embraces a lot of different life-ways). Even the Magdalenians, who left behind them such a prodigious record of creativity and symbolic behavior, presumably felt themselves to be integral parts of their surrounding ecosystems—although they may have had some role in the extinction of many large-bodied European mammals toward the end of the most recent Ice Age. But careless hunting to extinction of prey species, based as it may have been on the extraordinary human capacity, does not equate with any conscious attempt to change the environment. Certainly, we have no evidence that the Magdalenians were not content to live within the confines of the admittedly rich habitat that was their home. We may debate the meaning of Magdalenian art; but there can be little question that, whatever body of myth and belief that art represented, it incorporated at least implicitly the notion that these people were themselves an integral component of their natural habitat—which was thus a greater entity than they.

What a contrast with settled agriculturists! For the very concept of crop production implies the modification of ecosystems. As Eldredge notes, the adoption of agriculture involved, for the very first time, the abandonment of local ecosystems: "stepping outside," as he puts it. And once outside, humans could—indeed, virtually had to—view their place in nature differently; for indeed, in a very real sense, their place was now apart from nature. For an agriculturist, life is not a matter of cleverly exploiting what nature has to offer; rather, it becomes a battle with nature, a matter of sidestepping environmental vicissitudes through the application of technology. The battle might be lost or won, but it is a battle nonetheless, between two opposing forces.

Ironically, it seems that this new, confrontational way of life —and by implication new view of nature—may have been forced upon humans by nature itself. A drying trend followed the end of the most recent Ice Age (I shall not call it the last, for there are more to come) about 10 kyr ago. This made life difficult for hunter-gatherers in the Near East, who (as we know from the peculiar polish on their stone tools that this activity creates) were already harvesting wild cereals as part of their regular routine. It's reckoned that the Near Eastern landscape was already comparatively densely populated by humans at this stage, and moving to a more congenial habitat was probably not an option for these hunter-gatherers as they watched a major resource gradually dwindle in the hotter, drier summers. For beings with all the cognitive advantages of modern humans, it would not have been a major step to start planting seeds to replenish an important natural resource; and it appears that this is indeed what took place, mutant cereal varieties that shed their kernels less easily during harvesting being increasingly selected for sowing. Humankind had started on an inexorable cycle of intensifying artificial exploitation of formerly natural products.

Of course, humans being what they are, excessively intensive use of resources has almost inevitably resulted. Civilizations have risen and fallen on this cycle of overintensification, with population crashes resulting from economic crises due to mindless overexploitation (something we would do well to keep in mind in these days of skyrocketing human numbers). The point here, though, is that with increasing intensification comes increasing distance from the surrounding ecosystem and commensurate changes in attitudes toward nature. Eldredge finds evidence for this in the founding documents of the Judeo-Christian tradition, written by the descendants of those first hunter-gatherers turned agriculturists. Here, as he points out, we have firsthand accounts of how the early agriculturists saw themselves in relation to the natural world. Quoting Genesis 1:27 ("God said ... be fruitful, and multiply, and replenish the earth, and subdue it; and have dominion ... over every living thing that moveth upon the earth"), Eldredge characterizes this passage as "the most ringing declaration of independence ever set down." The independence to which he refers is, of course, independence from the rest of nature. We are the monarchs of the ecosystem in which we live; we have been given dominion over it by a greater power. But at the same time, it's important to note, the ecosystem is the enemy, to be "subdued." In practice, we don't automatically have dominion; we have to fight for it. Nature has to be tamed. And this adversarial relationship with nature has governed our attitude toward it right up to the present day, hardening our sense of entitlement in the process.

Perhaps, though, we should think twice about this attitude. We may indeed have stepped outside local ecosystems, but we are still dependent on them to an extent that we seem to be unwilling to acknowledge. The archaeological record shows how, over and over again, complex economic systems have collapsed following excessive population growth or the overexploitation and degradation of the local environment. Economists in our day tend to scoff at the notion that humans are exhausting the resources available to them; for in the recent history of our own civilization, at least, technology—human ingenuity—has always come to our rescue. The Green Revolution, for example, almost doubled grain production in some Asian countries at a stroke. And, of course, the economists have recent history on their side, at least so far. Thomas Malthus was certainly excessively gloomy, at least in the shorter term, when he predicted at the turn of the nineteenth century that human population growth would inevitably outstrip the food resources available; and the strident pronouncements of more recent doomsayers such as the Club of Rome (*The Limits to Growth*) have been proven dramatically wrong. Excessively specific predictions, after all, almost invariably are. Still, not a few local areas have already been abandoned by people because they have deteriorated to a point at which they can no longer be productively worked; and such unsettling phenomena as the apparently inexorable advance of the Sahara cannot be blamed entirely on natural causes.

More importantly, though, in this age of globalization it is the health of the overall world ecosystem that cannot be ignored. Nobody knows exactly how many people Earth can potentially support—estimates have varied from about five to twenty billion—but there's no doubt that the true number is finite and probably does lie within the upper limit of this range. Disaster may not immediately loom, but we should be uncomfortably aware that we have already exceeded the low end of the range of rational estimates. Nor is the matter of Earth's carrying capacity simply an economic one. There are social factors, too. Individuals need space; and limited space leads to social pathologies. We see this already in crowded cities and in overcrowded prisons; but the most harrowing glimpse of the future comes from monkeys.

In land-poor Japan, formerly free-ranging macaques have occasionally been maintained in extraordinary densities by artificial provisioning. When feeding time arrives, the monkeys descend in vast numbers upon their keepers. There is no social interaction among them, apart from conflict. Each individual monkey steadfastly occupies its tiny individual space and gobbles up pieces of food as fast as it can, staring intently at the ground where the food lies and avoiding the eyes of others in case aggression should result. What a far cry from the rich social lives of these primates when foraging under natural circumstances! The complex conflict-minimizing strategies of wild-living monkeys are nowhere to be seen; instead there is only this terrifying isolation of each cramped, nervous, and discomforted individual.

We have already, long ago, begun reaping the consequences of population growth, not simply economically, but socially, too. The maximum human community size in which standards of behavior can be maintained simply by social pressures seems to stand at about 150 individuals. Hunter-gatherer group sizes mostly lie well within this limit; but in settled societies it is soon exceeded. In an ideal larger society, individuals would get along with each other by voluntary compliance with social norms; but in practice, above this size limit, elaborate institutions—under which individuals frequently chafe and individual injustices abound—are needed to maintain society's cohesion. Few of us are able in these days to identify with a group as small as 150 people; and as a result, our sheer numbers seem to force us to seek our identities in smaller groups within the larger population. At one extreme, warfare between societies results; at the other, racism, religious prejudice, and similar unpleasant phenomena arise corrosively within societies. Often such destructive urges express themselves in the denial of "humanity" to one group by another, a phenomenon particularly well documented among soldiers, who are required to kill people whom they do not know and against whom they have no personal grievances. In our huge modern societies, with their inherently fissile tendencies, we are dangerously close to exhausting the social skills that are needed to knit those societies together.

Strolling past a church the other day, I noticed a sign that said: "Every day a child is born is a holy day," attributing this deep reflection to some wise person or other. If that person simply meant that we should live as if every day were a holy day, fine. But although the birth of each child is certainly a special event for those immediately involved, the daily arrival of unimaginable numbers of them is a recipe for future disaster as far as our species is concerned. The injunction of Genesis to "be fruitful and multiply" may have been reasonable in the context of the precarious economic existences of those brave early agriculturists (who certainly needed labor as their predecessors had never done); but it surely makes little sense now, at least as a prescription for our species as a whole.

Interestingly, we have recently seen a rise in fundamentalist environmentalism, which sensibly stresses the "stewardship" aspect of our relationship with the other components of God's creation (for if God gave us dominion, He surely also gave us responsibilities)—although this position is balanced on the opposite end of the fundamentalist spectrum by those whose attitude is "What do we care? The world's soon coming to an end anyway." The concern even of environmentalist fundamentalists does not, however, extend to matters of population control. After all, the scriptural admonition to multiply seems pretty clear-cut. Complicating the matter, too, is that in organisms such as ourselves, reproduction is intimately tied to sex—though in our days, the reverse is fortunately not true. And sex is one of those ancient limbic functions having little to do with the "higher" faculties that allow us to see ourselves as apart from nature. Indeed, at least at certain points in our lives, sex is probably the single most difficult element of our behavior for those higher faculties to control. The tension thus set up between our cortical and limbic structures is what leads to our society's ambivalence toward sex and to the consequent urge to regulate it. Tying sex to procreation is one way of achieving such regulation, and the paradoxical result is that attempts to control sexual activity (and to heed scriptural urgings) have led many in our society to advocate the effective decontrol of reproduction.

Well-intentioned though it may be, this is surely the ultimate example of human hubris at a time when global human population is turning from an economic asset into an enormous economic and ecological liability. Apart from nature we are, although only in a limited sense. And our stepping out of local ecosystems has served us reasonably well up to now. But what we cannot step out of is the global ecosystem. Our planet has been aptly described as an "oasis in space"; and that oasis, from which we have no plausible prospect of escape, is becoming increasingly despoiled by sheer pressure of human numbers. It is with the global ecosystem that our species will live or die; and if it is not to be the latter, we must find some way of abandoning, or at least of modifying, our arrogant and adversarial attitude toward the world that surrounds and supports us. Edouard Balladur, a former French prime minister, was presumably unaware of the full implications of his recent definition of civilization as the struggle against nature. Had he given the matter a little thought, he might well have been moved to declare that no civilization or society can long survive that shows too reckless a disregard for the natural world, from which our species' welfare can never be completely dissociated.

## *So: What Happened?*

In the opening pages of this chapter, I raised the question of human consciousness, that poorly understood quality that makes us feel so different from the rest of the living world. We cannot ignore our consciousness in any account of how we came to be what we are; but while the extraordinary reaches of this consciousness are unquestionably unique among animals, it is far from evident that only humans possess any kind of consciousness at all. Ultimately, the oft-heard claim that only humans have consciousness is no more than a matter of semantics, for all mammals are sentient, individualistic beings, not automatons. Structurally, their brains resemble ours in essential features; and if our brains are not machines, neither are theirs. Early in this book we surveyed the behavioral attributes of the apes and saw in them vivid echoes of ourselves. Having as we do such powerful feelings of being distinct from the other denizens of the living world, we prefer to emphasize the differences that we perceive between us and our closest living relatives, rather than the similarities; but in the final analysis, we have to admit that, anatomy apart, many of these differences are simply of degree—or sometimes, just possibly, not even differences at all.

In looking at what humans have become, I have drawn attention to the contradictory behaviors that stem at least in part from the untidy structure of our brains, the product of a very long evolutionary process of accretion and reorganization. I have done this because it is important to understand that our evolutionary past is always with us, even if some of our mental attributes are unprecedented. After all, we did not spring forth full-blown from the primordial ooze. Yet, as I've already emphasized, it's equally critical to realize that we are not behavioral prisoners of our genes in the way that evolutionary psychologists would have us believe. Even as it becomes almost daily clearer that many of our individual personality traits are genetically influenced, often very heavily, our emergent human capacities present us individually with an unparalleled menu of behavioral choices, the basis of our free will. And we make most of these choices for economic or social reasons that have nothing to do with spreading our genes. This having been said, however, it remains true that there's nothing essentially unique in the fact that our behaviors are not stereotyped. Some lions are cowardly, others brave; some dogs are smart (for dogs), some aren't; some chimpanzees are nicer than others. There can be little doubt that many other animals have individual consciousness in some form—any dog owner knows that; and, conversely, the example of sleepwalking demonstrates that quite complex human behaviors are possible in the absence of normal human consciousness.

When we look closely at those cognitive factors that make us what we are, we can often find parallels elsewhere in nature, and particularly among our primate relatives. They can anticipate; they can judge how to react; in a rudimentary way they can deceive; and they can experience contentment and attachment as well as fear and irritation. In the larger context of all those things that make up our everyday experience as individuals, the ways in which we differ from apes are in reality quite limited, at least in scope. Our brains, and the processes that go on within them, are not entirely different from those of apes; what has happened in the course of our evolution is that we have acquired new capacities—not in substitution for, but in addition to, our old ones. Exactly what is it, then, that lies behind our acquisition of fully human consciousness?

The key to this question lies, ironically, in the very difficulties we face in defining that consciousness. We can, in theory, discover what is unique about us both by comparing ourselves with the apes and by looking in our fossil and archaeological records for those changes that reflect our achievement of behavioral modernity. But in neither case is the endeavor at all simple, for a very significant reason: both sources of evidence are mute. Let's start with the apes. The reason why we cannot know the boundaries of their consciousness is a very straightforward one: they can't tell us. Apes can't talk. They can't explain to us what they are feeling or what is going on in their heads because they do not have language. And language is, as we have seen, not just a matter of stringing sounds together or even simply of communication. It is a complex business of making associations between stimuli received by the sensory parts of the brain and translating them into outputs from the centers that govern the production of sounds in the vocal tract. It involves categorizing and naming objects and sensations in the outside and inner worlds, making associations between them, and expressing the results according to an arbitrary set of rules. The key word here is *associations;* and although all sentient beings have some ability to make associations between different kinds of input into the brain, it's pretty clear that language not only depends on such associations, but also permits certain kinds of associations to be made. Language, it is fair to say, is more or less synonymous with symbolic thought.

This is not to say that all kinds of understanding are impossible in the absence of language and thought. As I've already noted, nonverbal intuition can tell us a lot about the world; and while it would not be accurate to view intuition as unexpressed speech, we can fairly regard it as a precursor, or at least as a precondition, of language. Pure intuition, however, has fairly dramatic limits. Intuition is individual and directly experiential; language is shared and symbolic. Wolfgang Köhler, the man who initiated studies of tool use among apes, once devised an experiment that vividly underlined the limitations imposed by the absence of language, even as it revealed understanding of a problem. He taught a pair of chimpanzees how to obtain a food reward by tugging simultaneously on two ropes hanging a dozen yards apart—too far apart for either chimpanzee to obtain the reward alone. Only concerted action by both would suffice. Once the pair had got the idea, one individual was removed and replaced by another who had not been taught the technique. The experienced chimpanzee clearly underwent fits of extreme frustration as the other failed to respond appropriately: he had a perfect intuitive understanding of the problem, but he could not carry out the task by himself and was simply unable to tell the other what to do to obtain the food. Of course, although he knew what to do, he had not arrived at the solution to the problem by rational thought; and it is vanishingly unlikely that he could have done so in the absence of the capacities that make language possible. For the abilities to arrive at the solution and to communicate it to others both stem from essentially the same mechanism.

Exactly what the cognitive correlates of language are is hard to specify, partly because of the problems of communicating with creatures that don't possess it. But it's as certain as anything inferential can be that language and the mental abilities directly associated with it loom large indeed behind the capacity to think, on which our species' success is founded. Language, like thought, involves forming and manipulating symbols in the mind; and our capacity for symbolic reasoning is virtually inconceivable in its absence. Imagination, too, is part of the same process; only once we create mental symbols can we combine them in new ways and ask, "What if?" Language is thus much more than simply the medium by which we explain our thoughts to ourselves and others: it is basic to the thought process itself. What's more, such vaunted human uniquenesses as anticipation of death—from which religious awareness and so much else flows—are probably little more than an extension of the same symboling capability. For however much we may fear it or be obsessed by it, our own death cannot be part of our individual experience. It remains an abstraction for each of us, no matter how many symbolic deaths of others we may see nightly on TV and movie screens. The same essential capacity also lies behind the "inner eye" that allows the brain to objectify its own workings and behind the inner eye's equally unusual alter ego, self-deception, both of which similarly depend on the ability to generate and recombine mental symbols. And on and on: virtually any component of our ratiocinative capacities you can name—from our sense of humor to our ability to entertain apocalyptic visions—is based on those same mental abilities that permit us to generate language.

Our fossil record shows pretty clearly that our ancient precursors were nonlinguistic. As we've seen, we cannot read language abilities directly from the external form of brain casts; but the structures of the vocal tract that make articulate speech possible imprint themselves on the base of the skull. The earliest stone toolmakers certainly had a clear insight into the mechanical properties of various rock types and were able to anticipate future needs; but this intuitive understanding was achieved in the absence of the peripheral structures necessary for speech, which make spoken language possible. Even such later innovations as the control of fire and the building of shelters appear to have been achieved purely through refinement of the capacity for intuitive insight. *Homo heidelbergensis* is the first hominid we know of that had a skull base designed to accommodate a vocal tract of modern kind. These human precursors were efficient stone toolmakers, fire users, and shelter builders, and they had pretty big brains to boot; yet the behavioral record they left behind contains virtually no evidence of symbolic activity. The archaeological record is, of course, selective; but the contrast here with the symbol-strewn Upper Paleolithic could hardly be more stark. Almost certainly, then, the descent of the larynx and the elaboration of the supralaryngeal structures that underlie speech occurred in a more general, respiratory context. And it follows from this that the modern vocal tract arose as an exaptation for language: a preexisting condition that made this remarkable innovation possible once the necessary brain wiring had been acquired.

It is only with the arrival of Cro-Magnons in Europe (together with tantalizing earlier hints of behavioral modernity in other parts of the world, perhaps most significantly Africa) that the archaeological record unequivocally declares that people of modern sensibility had arrived. And it's surely no coincidence that it is only with the Cro-Magnons and their like that archaeologists confidently perceive evidence of economic strategies that resemble those of modern hunting-gathering peoples and that reflect the uniquely human capacity to read and understand the external environment. Even the Neanderthals, complex and admirable as they may have been, were probably limited to an intuitive level of understanding of the world about them. Neanderthals were expert craftsmen in stone and were adept at learning by imitation, even if they individually showed rather little originality and inventiveness. They were evidently capable of empathy and may even have wept as they interred that body on its bed of flowers at Shanidar; and there can be little doubt that they had quite sophisticated ways of communicating with each other. All this, however, was in every likelihood achieved in the absence of symbolic reasoning—and perhaps even of the latent ability for articulate speech, as the alleged curiously primitive skull-base anatomy of some Neanderthals, at least, suggests. The Neanderthals did not live, as we do, in a world of their own making, reconstructed in their minds, but in the world as nature presented it to them.

Their archaeological record is extensive enough for us to be certain that the Neanderthals were not our intellectual equals. But it would be profoundly inaccurate to construe them simply as an inferior version of ourselves. Given our inability to imagine states of consciousness other than our own, it might be tempting to compare Neanderthal intelligence with that of the stupidest, most unimaginative, most humorless, most inarticulate people you could conceive of today. It would, however, be profoundly wrong. Neanderthals, for all of their anatomical and behavioral similarities to us, were beings of a different order and need to be understood on their own terms. Almost certainly, they lacked the crucial capacity for symbolic thought of which language is emblematic and of which even the most limited modern intellect is capable. We can, then, most usefully view the behavioral record bequeathed to us by the Neanderthals not as an inferior reflection of our own ways of doing business, but as a measure of the—altogether remarkable—achievements that can be made purely through the workings of intuitive insight.

How and when, then, did we acquire our singular linguistic/symbolic abilities? As I've remarked, for all our extensive knowledge of how the human brain is put together, how information flows through it, and how certain parts function in particular behaviors, the mechanisms by which our complex symbolic consciousness is generated remain totally obscure. What is clear, however, is that after a couple of million years of erratic brain expansion and other acquisitions in the human lineage, the necessary exaptations must have been in place to permit the completion of the whole extraordinary edifice by a change that was presumably rather minor in genetic terms. Just as the keystone of an arch is a tiny portion of the whole structure yet is vital to its integrity, a relatively small neural change must have had this remarkable emergent effect in our brains. And this neural innovation must, of course, have been acquired in a small population of our ancestors at a time when any essential peripheral equipment—the vocal apparatus, for example—was already available to allow its expression.

Nobody yet understands exactly why brain enlargement and elaboration has been such a consistent, if episodic, theme over the long evolutionary history of humans—indeed, of primates in general. Neither do we know why, at the end of this process, the human brain had become so beautifully exapted for language and symbolic reasoning. Perhaps the advantages conferred by improvements in intuitive reasoning were by themselves sufficient to carry this process along, although there must certainly have been something special about early hominids to make it so aggressively characteristic of our group. What is undeniable, however, is that exaptation is nothing very special in itself. For example, it has recently been shown that speaking and writing abilities are located in opposite halves of the brain. The ability to write is thus not a simple passive consequence of the mental capacity to form spoken words; it depends on distinct neural circuits. The latent capacity for both behaviors was certainly present in the early modern human brain; yet the ability to write was not discovered until 50 kyr or so after the invention of spoken language—and among many groups was not discovered at all.

Unhappily, our fossil and archaeological records are frustratingly incomplete for the critical period in our evolution during which the modern human capacity emerged. The evidence is strong that humans of modern anatomy arose in Africa; but for the moment at least it appears that this was an early event compared to any indications we have of complex modern symbolic behaviors. As I have emphasized, there is no reason to suppose that the appearance of new technologies and behaviors need necessarily go hand in hand with the arrival of new species; but it's still important to know how *Homo neanderthalensis* managed to coexist in the Levant for upward of 60 kyr with Mousterians who in their bony anatomies were plainly *Homo sapiens*. Does this long period of coexistence—and cultural similarity—mean that the brains of these hominids were effectively identical? For, if so, the internal hard wiring of the modern brain must somehow have been acquired many millennia subsequent to our acquisition of modern skeletal structure—as has, indeed, been suggested. Most of the time I find this notion a little hard to believe. For if it is true, behaviorally modern humans acquired the anatomical basis of their symboling capacities rather late. This would not have allowed creatures of this new physical kind much time to spread from their putatively African center of origin and to replace, worldwide, people who looked just like them skeletally but who had more archaic (if externally identical) brains. What's more, there's absolutely no evidence in the admittedly poor record to support this scenario.

The only evident alternative is that modern human bony anatomy arrived as a package with the exapted modern human brain and that this remarkable new organ lay fallow, as it were, until some cultural stimulus—almost certainly the invention of language—kicked it into action in one local population. It's even possible that the human capacity originated at least partly in an epigenetic, developmental event, rather than in a major change in the genetically programmed structure of the brain. Development of the brain after birth includes the creation of specific pathways out of a complex mass of neural interconnections, largely through selective elimination. At an early stage of life, languages are easily learned, even several of them simultanously; after the age of ten, the language-specific pathways have become established, and the acquisition of new languages is much harder. Language capacities thus involve the developmental structuring of the brain, and it is not too hard to envisage in broad terms how an initial, relatively rudimentary form of language could have been acquired—perhaps initially even among children—by developmental means, subsequently refining itself and diversifying over the millennia to produce the bewildering variety of languages existing today. Certainly, the emergence of language as it is universally familiar to us today cannot have been an overnight event. Whatever the case, it is surely a great deal more plausible to envisage that language, with its associated mental abilities and behavioral complexities, spread (and diversified) from its place of origin through contact and diffusion among established human populations that already possessed the latent ability to acquire it, than that numerous populations of people who looked just like us, but who lacked that ability, were eliminated worldwide over a relatively short period of time.

It is frustrating indeed to come to the end of our story and to have to admit that we have little idea as to exactly how, when, where, or why our extraordinary consciousness was acquired. However much we tend to be obsessed by them, our cognitive capacities, epitomized by our linguistic abilities, do indeed mark us off distinctly from all of the millions of other creatures on the planet. But the latent ability to form and manipulate mental symbols is clearly not the predestined result of an inexorable process over the aeons, even if the foundations for it were established over a long human evolutionary past. Rather, its acquisition was an emergent event that was probably rather minor in terms of physical or genetic innovation, that was comparatively sudden, and that came very late in our evolutionary history. This event of events is, alas, probably undetectable from the bones and teeth that reveal our fossil history; and the archaeological record, as we've seen, is incomplete, highly selective, and but a dim mirror of the behavior of our forebears. But although the initial probability that all the components needed for modern human consciousness would come together precisely as they did was undoubtedly minuscule in statistical terms, so was the probability of *any* of the millions of specific outcomes of the evolutionary process. Viewed this way, the event itself is far less remarkable than its end product.

What, then, are we to make of ourselves? Well over three billion years after life established itself on Earth, we, alone among the millions of descendants of our ancient common ancestor, somehow acquired not just a large brain—the Neanderthals had that—but a fully developed mind. This mind is a complex thing, not in the sense that an engineered machine is, with many separate parts working smoothly together in pursuit of a single goal, but in the sense that it is a product of ancient reflexive and emotional components, overlain by a veneer of reason. The human mind is thus not an entirely rational entity, but rather one that is still conditioned by the long evolutionary history of the brain from which it emerges. Great though may be the leap we have made away from the rest of the living world in the acquisition of symbolic thought, we have not entirely emancipated ourselves from the brain structures that governed the behavior of some very remote ancestors indeed. And it is precisely this interaction of the ancient with the new that makes us not only unique in many very admirable ways, but also uniquely dangerous—as much to ourselves as to the rest of the living world.

Because the fossil and archaeological records show us that the final step in becoming human was more than a simple extrapolation of earlier trends, it is hardly for a paleontologist to attempt an explanation of the complexities of modern human behavior, at least beyond pointing out the interaction of the old and the new that goes on inside our symbol-manipulating brains and that underpins our vaunted consciousness. Exegesis of the way we are lies in the domains of psychologists, neurobiologists, philosophers, novelists, playwrights, and others. How we ought to conduct our lives (and how we are encouraged to live them) is the province of ethicists, philosophers, and, God help us, lawmakers. And the practical fate of humanity lies in the hands of those same politicians and billions of ordinary people. But a paleontologist can nonetheless point out that nature, while having placed a unique and potentially highly destructive capacity in the hands of *Homo sapiens*, is under no obligation to ensure we use that capacity wisely. Still, since it's undeniable that we are the product of a long process of change, the question inevitably arises of whether further change is on the cards that might actually help us deserve our zoological name of "Man the Wise." So, what *can* we expect of our evolutionary future? That's the subject of the concluding pages of this book.

. . . . . . . . . . . . . .

# *Postscript*

People are perennially interested not only in where they came from, but in where they are going—and, indeed, ever since the notion of evolution emerged in the last century, our origin and destiny have somehow seemed inextricably interlinked. Given the prevailing view of evolution as a gradual, progressive process, it's not difficult to understand how we so readily make this connection; and the rapidity in our own times of cultural change (transmitted, of course, in a totally different way from evolutionary change) has superficially bolstered this view. It's significant also that the booming business of futurology has always been one of trend spotting and extrapolation. The past has never had any problem in selling itself as the key to the future; where true visionaries have traditionally had trouble is in getting the totally unfamiliar accepted.

No wonder, then, that popular views of where our species is headed frequently present visions of creatures with big, brainy, supremely rational heads. These heads are precariously perched on scrawny, virtually limbless bodies, while their possessors luxuriate amid vistas of unending technological miracles. As extrapolation this is fairly reasonable; for it is undeniably the case that over the long haul, the human brain has become larger, that lately our bodies have become modestly less robust, and that our technology has become more complex—in recent decades with extraordinary rapidity. But can we really conclude that our remarkable past performance is any guarantee of future improvement? I deeply doubt it.

To start with, it's clear that evolutionary change in our past has been sporadic at best. Indeed, there was no change of major functional consequence over the entire first half of our evolutionary history. Even the first stone toolmakers did not, apparently, look very different from their predecessors; and once they had arrived, another million years were to pass before another significant technological improvement was made. And on and on. In both the anatomical and the technological realms, the history of our lineage has been one of episodic innovation, and not one of a gradual approach to perfection.

Disappointing though this may be, it fits pretty well with our emerging appreciation of the complexities of the evolutionary process. If evolution involves diversification and speciation rather than the steady accumulation of small changes, and the history of individual species is for the most part one of business as usual, then we should not be surprised to find that significant innovation in our own lineage has been an intermittent phenomenon. Thus, whether we wish to call upon evolutionary theory or upon our own fossil past as a guide to what we might expect the biological future of our species to be, the question we must ask is the same: Do the conditions now exist for new evolutionary developments out of *Homo sapiens?*

We have seen that it is small isolated or quasi-isolated populations that are both the true engines of innovation in evolution and the target of speciation. Such populations of early humans were widely scattered around the globe during the ecological and geographical disruptions of the Ice Ages, and it was in this context that *Homo sapiens* emerged. Today, however, the situation is dramatically different. The worldwide human population is well over five billion and mushrooming. Modern transportation has made a mockery of distance and geographic barriers. Individuals are incomparably more mobile than ever before. Isolation of populations within our species is a thing of the past, while intermixing among peoples from all geographical areas is on the upswing. The result, without question, is that today the conditions for true evolutionary innovation within our species simply do not exist. We form a vast single population that straddles Earth; never ever, in the history of any species, have conditions been less propitious for the fixation of evolutionary novelties. And, short of some calamity, they won't be.

Of course, calamity may not be very far away. Some equally appalling but less fragile and more easily communicable virus than HIV may be waiting around the corner. Technology could probably do rather little to mitigate the worldwide consequences of a meteorite impact of the kind associated with the demise of the dinosaurs. The latest estimate puts the probability of such an event at close to zero for the next one hundred thousand years; but an eventual impact is inevitable. More immediately, our snowballing high-tech wizardry might well outstrip our ability to control one or more of its myriad consequences. Equally worrying is the likelihood that excessively intensive exploitation of the worldwide ecosystem may lead to its—and our—collapse. The possibilities are endless, and any agency that might work to reduce and fragment human populations would reinstate the possibility of human evolutionary change.

Calamity aside, though, does any of this matter? Well, in a way it may, at least to the extent that the science fiction writers' fantasies reflect some kind of yearning in our collective unconscious. It's obvious, for example, that the creators of *Star Trek*'s Spock hit close to home when they invented that supremely rational being. For there is undeniably something bizarrely seductive in the notion of living unencumbered by all of the emotional baggage that being human inevitably entails. After all, irrational behavior, propelled by obscure emotion, has lain behind most of the endless misery that the human species has inflicted on itself (and others) over the course of recorded history. Yet there's a flip side to this, of course, for none of us would wish to be an automaton, and a life without emotion would be a life without exhilaration, love, and joy.

Sci-fi possibilities for our future do not stop with Spocks, though. One possibility for change that is increasingly mooted is offered by space travel. What if small human colonies were to be established on far-off celestial bodies? Population sizes would be small, and the conditions for evolutionary change would be in place. Or would they? Even if space travel of this kind were eventually to become a reality, which is far indeed from certain, any such colony would have to be sustained by a lifeline from Earth. Isolation would not be complete, and the evolution of a new species would be vanishingly unlikely. And even if by some miracle a space colony were able to survive in isolation, if its lifeline were cut, any subsequent developments would be biologically irrelevant to continuing human life on Earth. Similarly with genetic engineering, another sci-fi favorite for future genetic and evolutionary change. New, artificially produced genotypes could only be sustained by sequestering individuals possessing them from the population at large. In the unlikely event that it were ever deemed permissible to carry out such manipulations on members of our own species, genetically engineered innovations would necessarily remain uniquely the property of small "laboratory" populations and thus would fail to affect the much larger *Homo sapiens* population as a whole.

So the news, then, is both good and bad. The bad part is that if things persist more or less as they are, we can't expect evolution (or technology) to ride in on a white horse, as it were, and rescue the human species from its follies by endowing us with overweening intelligence or even collective good sense. The good news, on the other hand, is that barring disaster, we will almost certainly forever be the idiosyncratic, unfathomable, and interesting creatures we have always been. Short of the unthinkable, we are stuck with our old familiar—and potentially dangerous—selves, and we urgently need to learn how best to live with that fact. Perfectibility remains, as ever, an illusion.

## FURTHER READING

There is a vast literature covering the themes addressed in this book. Following are a few easily accessible recent works, listed according to the chapter in this volume to which they pertain. Most of them contain ample references to the more specialized literature.

### *Chapter 1*

Bahn, P., and J. Vertut. *Images of the Ice Age.* New York: Facts on File, 1988.

Marshack, A. *The Roots of Civilization.* Rev. ed. Mount Kisco, NY: Moyer Bell Ltd., 1991.

Pfeiffer, J. E. *The Creative Explosion: An Inquiry into the Origins of Art and Religion.* New York: Harper & Row, 1982.

Sieveking, A. *The Cave Artists.* London: Thames and Hudson, 1979.

Tattersall, I. *The Human Odyssey: Four Million Years of Human Evolution.* New York: Prentice-Hall, 1993.

White, R. *Dark Caves, Bright Visions: Life in Ice Age Europe.* New York: American Museum of Natural History and W. W. Norton, 1986.

### *Chapter 2*

Ackerman, S. *Discovering the Brain.* Washington, D.C.: National Academy Press, 1992.

Byrne, R. *The Thinking Ape: Evolutionary Origins of Intelligence.* Oxford: Oxford University Press, 1995.

Corballis, M. C. *The Lopsided Ape: Evolution of the Generative Mind.* New York: Oxford University Press, 1991.

Goodall, J. *The Chimpanzees of Gombe: Patterns of Behavior.* Cambridge, MA: Belknap Press, 1986.

McGrew, W. B. *Chimpanzee Material Culture: Implications for Human Evolution.* Cambridge: Cambridge University Press, 1992.

Pinker, S. *The Language Instinct.* New York: William Morrow & Co., 1994.

Savage-Rumbaugh, S., and R. Lewin. *Kanzi: Ape at the Brink of the Human Mind.* New York: John Wiley & Sons, 1994.

Wallman, J. *Aping Language.* Cambridge: Cambridge University Press, 1992.

## *Chapter 3*

Dawkins, R. *The Extended Phenotype: The Long Reach of the Gene.* New York: W. H. Freeman & Company, 1982.

Eldredge, N. *Reinventing Darwin: The Great Debate at the High Table of Evolutionary Theory.* New York: John Wiley & Sons, 1995.

———. *Time Frames: The Rethinking of Darwinian Evolution and the Theory of Punctuated Equilibria.* New York: Simon & Schuster, 1985.

———. *Unfinished Synthesis: Biological Hierarchies and Modern Evolutionary Thought.* New York: Oxford University Press, 1985.

Ridley, M. *Evolution.* Boston: Blackwell Scientific Publications, 1993.

## *Chapters 4 and 5*

Johanson, D. C., L. Johanson, and B. Edgar. *Ancestors: In Search of Human Origins.* New York: Villard Books, 1994.

Leakey, R., and R. Lewin. *Origins Reconsidered: In Search of What Makes Us Human.* New York: Doubleday, 1992.

Lewin, R. *The Origin of Modern Humans.* New York: Scientific American Library, 1993.

Schick, K. D., and N. Toth. *Making Silent Stones Speak: Human Evolution and the Dawn of Technology.* New York: Simon & Schuster, 1993.

Tattersall, I. *The Fossil Trail: How We Know What We Think We Know about Human Evolution.* New York: Oxford University Press, 1995.

———. *The Last Neanderthal: The Origin, Success, and Mysterious Extinction of Our Closest Human Relative.* New York: Macmillan, 1995.

Tattersall, I., E. Delson, and J. A. Van Couvering, eds. *Encyclopedia of Human Evolution and Prehistory.* New York: Garland Publishing, 1988.

## *Chapter 6*

Allman, W. F. *The Stone Age Present: How Evolution Has Shaped Modern Life—From Sex, Violence, and Language to Emotions, Morals, and Communities.* New York: Simon & Schuster, 1994.

Eldredge, N. *Dominion: Can Nature and Culture Co-Exist?* New York: Henry Holt & Company, 1995.

Humphrey, N. *Consciousness Regained: Chapters in the Development of Mind.* New York: Oxford University Press, 1983.

———. "The Uses of Consciousness." *57th James Arthur Lecture on the Evolution of the Human Brain.* New York: American Museum of Natural History, 1987.

Wright, R. *The Moral Animal: Why We Are the Way We Are: The New Science of Evolutionary Psychology.* New York: Pantheon Books, 1994.

# INDEX